Automation Made Easy

Everything You
Wanted to Know
about Automation—
and Need to Ask

AF173196

Automation Made Easy

Everything You
Wanted to Know
about Automation—
and Need to Ask

Peter G. Martin and Gregory Hale

Automation Made Easy: Everything You Wanted to Know about Automation—and Need to Ask

Copyright © 2010 by   International Society of Automation
                        67 Alexander Drive
                        P.O. Box 12277
                        Research Triangle Park, NC 27709

ISBN: 978-1-936007-06-6

Notice
The information presented in this publication is for the general education of the reader. Because neither the author nor the publisher has any control over the use of the information by the reader, both the author and the publisher disclaim any and all liability of any kind arising out of such use. The reader is expected to exercise sound professional judgment in using any of the information presented in a particular application. Additionally, neither the author nor the publisher have investigated or considered the effect of any patents on the ability of the reader to use any of the information in a particular application. The reader is responsible for reviewing any possible patents that may affect any particular use of the information presented. Any references to commercial products in the work are cited as examples only. Neither the author nor the publisher endorses any referenced commercial product. Any trademarks or tradenames referenced belong to the respective owner of the mark or name. Neither the author nor the publisher makes any representation regarding the availability of any referenced commercial product at any time. The manufacturer's instructions on use of any commercial product must be followed at all times, even if in conflict with the information in this publication.

Library of Congress Cataloging-in-Publication Data

Martin, Peter, 1952-
  Automation made easy : everything you wanted to know about automation
and need to ask / Peter G. Martin and Gregory Hale.
      p. cm.
  Includes bibliographical references.
  ISBN 978-1-936007-06-6 (pbk.)
  1. Manufacturing processes--Automation. 2. Industrial
management--Automation. 3. Automation. I. Hale, Gregory. II. Title.
  TS183.M3656 2009
  670.42'7--dc22
                        2009023231

I dedicate this book to my best friend, love, and wife – Liz, and to my children, Derek and Erin, their spouses, Jennifer and David, and my beautiful granddaughter Karly. I love you all!
—Peter Martin

There comes a time when a person understands you just don't want to go through life alone. I came to that realization almost 30 years ago when I met my wife, Sharon, and that sense became even stronger with my children, Lindsay and Chris. While I seriously doubt they will ever read it, I dedicate this book to Sharon, Lindsay, and Chris. In their own inimitable way, they make it easy for me to strive to be a better person every day. Thank you.
—Greg (a.k.a. Dad)

Contents

About the Authors

Peter Martin has worked in industrial automation for more than 30 years in a variety of roles, primarily with Invensys, although he also worked at Automation Research Corporation (ARC). He was named one of the 50 most influential innovators in control and automation by *InTech* magazine and as a Hero of U.S. Manufacturing by *Fortune* for his work in real-time performance measurement and management, for which he also holds multiple patents. He has authored two other books, *Bottom-Line Automation* (ISA) and *Dynamic Performance Management: A Pathway to World Class Manufacturing* (Van Nostrand Reinhold), and is a contributing author to *A Guide to the Automation Body of Knowledge* (ISA) and *Metrics that Matter* (MESA). Dr. Martin has also published numerous papers and articles in the area of industrial automation. He holds a BA and an MS in Mathematics, an MA in Administration and Management, a Master of Biblical Studies, a Doctor of Engineering in Industrial Engineering, and a Ph.D in Biblical Studies.

Gregory Hale is the editor of *InTech* magazine, the official publication of the International Society of Automation. Prior to starting at *InTech* in 1999, Hale was the editor in chief at Post-Newsweek's *Reseller Management* magazine. His extensive editorial, web, and management experience includes executive editorial positions with CMP's *Computer Reseller News* and *Tour & Travel News*, the *Times Herald-Record* (Middletown, NY), *The Knickerbocker News* (Albany, NY), *The Boston Globe*, and various other New England newspapers.

Acknowledgments

Thank you to Greg Hale, who encouraged me to work with him on this project for a number of years. Greg has a great ability for taking technical gibberish and converting it into an understandable, interesting, and fun presentation. Greg is an industry icon who impacts industrial automation and everyone he works with in a very positive way, every single day.

I also want to thank ISA's Susan Colwell, who continually encouraged us and invested huge amounts of time and effort in helping direct the manuscript through its various stages.

A number of people I work with at Invensys have helped move this project forward in a variety of ways. They include Melanie Russell, Mary Beth Connolly, Mark Davidson, Jim Duckels, Sherry Allphin, Paulett Eberhard, Neil Cooper, Phil Clark, Sudipta Bhattacharya, Bob Cook, Lauri Crawford, Don Clark, Russ Barr, Bob Cook, Carol Vallett, John Eva, Rick Garlock, Jen Linell, Vicki VanDenBerghe, Kevin Fitzgerald, Steve Tiller, Tim Sowell, Grant Le Sueur, Rashesh Mody, Pankaj Mody, Steve Young, and Jack Clarke. All of the students, who participated in the course that resulted in the writing of this book, have my thanks.

I also want to thank my wife, Liz, for supporting me through this effort and the rest of my family for supporting me even when they did not know they were.

–Peter Martin

First and foremost, a big thank you to Peter Martin for allowing me the privilege of working with him on this book. Among his vast array of talents, Peter knows the industry through and through and is not afraid of stepping outside the "normal" boundaries of traditional thinking.

I also want to thank ISA's book publishing maven, Susan Colwell, for being a true champion and taskmaster in pushing this book through all the channels.

Recognition and thanks are due to *InTech* staffers Nick Sheble, Ellen Fussell Policastro, and Emily Kovac for being great information sources and sounding boards. In addition, kudos to former *InTech* staffers George Davis and Jim Strothman for their continued support and allowing me to bounce ideas off them.

In addition to my wife Sharon and my children Lindsay and Chris, heartfelt thanks go to the rest of my family: my mother Vesta, my late father Thomas, my sister Pat and brother-in-law Tom, my brother Bob and my other brother Ron and his wife Marcia, and my late brother Dave. Being the youngest of this crew was truly a great learning experience, one that would be impossible to forget. I also want to thank my extended family: Jim and Pat, Debbie and Rich, and the late Fran and Frank. They are the greatest.
                                                                              –Greg Hale

Preface

You see it all the time: a new head of a department, or of a company for that matter, comes into a new position in the industrial automation environment from a different industry. The press release talks in glowing terms of the skills and capabilities the new office holder had in his or her previous position. This person "will leverage his strengths and capabilities to help usher in a new era." They have to hit the ground running and start producing yesterday.

However, there is a catch. Over the past fifty years, the field of industrial automation has evolved from a number of independent technical fields, such as instrumentation, electronics, maintenance, plant operations and computer science. These traditionally independent fields converged to form today's industrial automation. This convergence contributes to making the study of industrial automation much more confusing than it really should be. Part of the reason for this is that any technology-based area of study has its own idiosyncratic terminology, jargon, and slang, including acronyms. This can provide a huge barrier to developing a functional understanding for anyone walking into the industry. As a mathematics professor proclaimed to a class finishing their first year of graduate school, "We spent the first year learning the words and now it's time to learn some math." There is more truth to this than any of us might like to admit.

When the digital computer showed promise as a tool to solve industrial automation problems, the lexicon of computer technology merged with the lexicon of pneumatic and electronic instrumentation and control systems. It was not surprising that instrument companies divided into two groups, the instrument and control experts and the computer gurus. These two groups had great difficulty talking to each other because they did not have a common language. Sometimes the same acronym had two very different meanings. SPC to the computer professionals in automation companies may have meant set point control while to the operationally focused team it meant statistical process control.

To exacerbate this situation, a considerable amount of the technology and terminology associated with industrial automation comes from digital computer markets and technologies. The reason this adds a level of confusion is that computer science is one technical field in which the terminology is not driven by academics; rather it is driven by marketing departments. Digital Equipment Corporation introduced their Programmable Data Processor (PDP) series of computers to the marketplace a number of years ago as the world's first minicomputer. The word minicomputer became a part of Digital's marketing campaign. Digital intentionally did not define this word because by not defining it they could more easily claim, without having to technically justify their position, that competing computer companies did not really make a minicomputer. College professors spent the next twenty years trying to develop a technical definition for "minicomputer," and to the best of the authors' knowledge they were never truly successful. This characteristic of the lexicon has resulted in a set of words and phrases in industrial automation like distributed control system (DCS), programmable logic controller (PLC), and manufacturing execution system (MES) to name only a few; terms used daily, but not well defined.

There are aspects of industrial automation based on rich and deep technology that require considerable in-depth study to understand them, but from a functional perspective most of industrial automation is pretty straightforward. The catch is that the field is dominated by technologists who cannot help but try explaining relatively simple issues in excruciating technical detail, causing many an eye to glaze over.

There are a slew of books that provide detailed explanations of each of the major aspects of industrial automation, including all the mathematics and formulas and dynamic models. These are important books for those getting into the heavy detail. But the end result is that without a preliminary understanding of industrial automation, these are very difficult to comprehend.

Our purpose in writing this book is to provide a basic functional understanding of industrial automation. It has been very tempting to delve into technical details in a number of topics, but that is not what this book is all about.

There are people moving into industrial automation as part of their professional development. That movement includes, but is not limited to, executives who have come into industrial automation after leading companies in other markets. That level of change is good for industrial automation as new approaches and ideas often accompany new talent. We find that there are many people becoming associated with the world of industrial automation from a variety of other related disciplines, such as information technology or accounting, who require a basic level of understanding of automation to perform their job functions more effectively. These people who are new to this field need a way to quickly educate themselves with the technology and terminology; time is of the essence. We hope this book fills that need.

We have structured the material in this book to progress from the most basic subject matter through more advanced automation topics. Depending on your background and level of exposure to manufacturing processes and automation, you may want to consider skipping over some of the earlier chapters and proceed directly to the chapters of prime interest.

CHAPTER 1

Manufacturing and Production Processes: The Raw Facts

When it really comes down to it, companies exist to earn a profit. Manufacturers are no different. Simply put, manufacturing is the making or processing of raw material into finished products, especially by a large-scale industrial operation. Before discussing the three different types of manufacturing processes, let's take a quick look at the basic components and characteristics common to every manufacturing process.

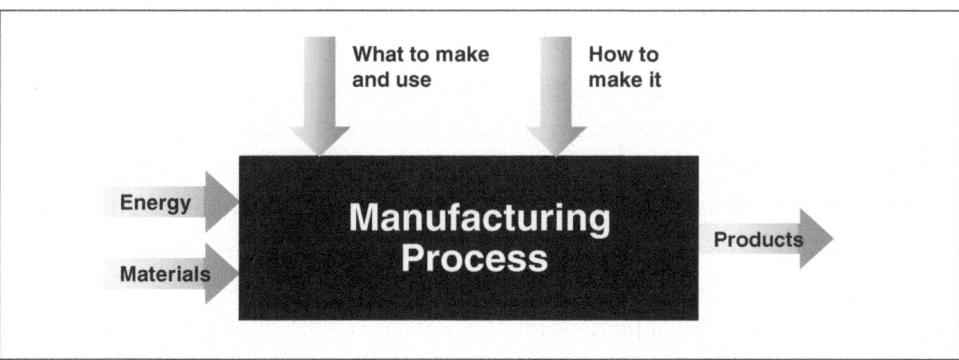

Figure 1-1 General Manufacturing and Production Process

Every manufacturing process is designed to transform raw materials into products through the utilization of basic production resources, such as equipment, tools, energy, and manpower. Figure 1-1 shows that the primary inputs to a manufacturing process include energy and raw materials. The primary output of a manufacturing process is one or more products or grades of product.

Gasoline is a perfect case in point. A quick snapshot of the process (a more in-depth version is discussed below) has crude oil coming to a plant after being pumped out of the ground. It then goes through a complex heating and cooling process where one of the end results is gasoline, which you use to fill up your SUV.

In multiple-product manufacturing operations, manufacturers often need to make decisions on what product they want to make at any point in time, and if there are multiple options within the process, how they should make the product. These two functions are scheduling and production planning. These basic components and concepts hold for any manufacturing process.

Three types of manufacturing processes exist: continuous, batch, and discrete. All three of these manufacturing process types have the basic characteristics discussed above, although they are very different in key aspects of their operation. These processes are not mutually exclusive as there are manufacturing operations that include all three types, although operating only one of the three is common. Manufacturing professionals often refer to the operation of a plant according to the dominant manufacturing process type employed. For example, an oil refinery may be referred to as a continuous process plant, even though there may be other types of processes going on at the plant.

Continuous Processes

"Continuous" simply means a manufacturing process where raw materials and energy are consumed in a continuous stream, and a product results. That product continues to be made in an ongoing manner once the process starts. Take, for example, the float glass process (Figure 1-2). Sand and other ingredients continuously feed into a large furnace. After the raw materials melt, they flow onto a molten metallic bed, where they form a sheet. After being formed, the molten glass sheet is allowed to cool slowly, and as it cools, it hardens into a continuous plate of glass that is then annealed to prevent internal stress and finally cut into sections. Once a float glass plant starts up, it typically operates continuously for years.

Products produced via a continuous manufacturing process typically do not have to be made this way. They can also be made in a discontinuous manner. The production of plate glass is a perfect example. In the eleventh century, manufacturers made glass panes one at a time using a glass blowing process with a flattening process. Although this process worked quite well, it was very limited in terms of the amount of glass a manufacturer could produce. A float glass plant operating in a continuous manner can produce much more glass than a discontinuous glass-making

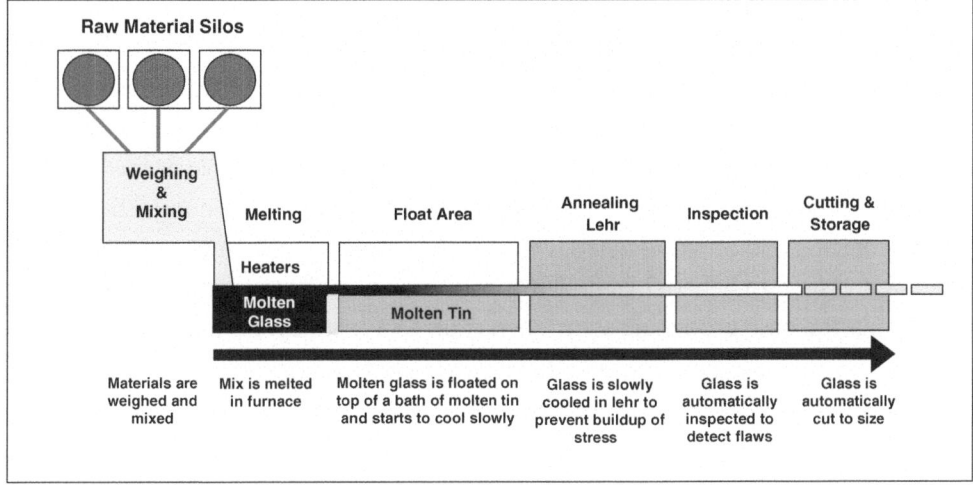

Figure 1-2 Float Glass Plant

process can. Continuous processes increase the level of production a manufacturer can achieve.

Continuous processes make the most sense when the market demand for the product is high, and the output of the manufacturing process has to be equally high in order to meet the demand. Therefore, it is important to understand that designing a continuous process to manufacture products is a decision based on big market demand for the product. Gasoline is a good example.

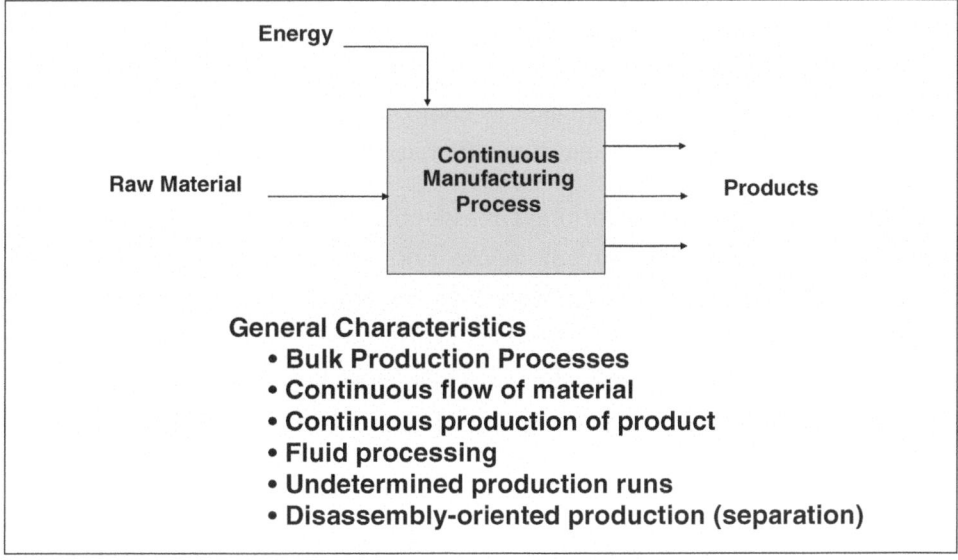

Figure 1-3 General Characterization of Continuous Processes

There are a number of characteristics typical of continuous processes (Figure 1-3). One is that most continuous processes are fluid-based. That is, they involve a significant amount of either liquids or gases as raw materials or intermediates in the processing. In order to make product on a continuous basis, there must be a continuous amount of materials on hand, which are naturally available as fluids. The glass plant example is interesting in this respect. It starts with a continuous charge of a mix of sand and other materials, which, although solid, are composed of small particles and tend to behave in a fluid manner. This mix melts to form a true fluid throughout the base processing steps, then cools to a solid toward the end of the process. This is a continuous process that involves solids, liquids and gases, but it behaves, for the most part, as a fluid process.

Another characteristic of continuous processes is that they tend to have undetermined (open-ended) production runs. As was previously discussed, float glass processes may have production runs that are measured in years. The same is true for oil refineries, which are largely continuous processes. Continuous processes tend to be challenging and take considerable time to start up, so manufacturers want them to operate as long as possible. It is a simple business formula: the more time they operate, the more product they make. The more product they make, the more revenue they generate.

Continuous processes are, for the most part, invisible. This is because much of fluid processing takes place within pipes and vessels, out of the view of operators. Therefore, these processes typically require at least a low level of automation in the form of instrumentation, just to be able to operate effectively. Batch and discrete processes, on the other hand, can often operate quite effectively in manual mode, with no need for any level of automation technology, although there are some complex batch processes that require automation to operate effectively.

Most people envision manufacturing as assembling a final product from a number of subunits, such as assembling an automobile. Continuous processing is often a reverse approach, although there are some assembly-based continuous processes, such as continuous chemical reaction and float glass manufacturing. Often a single feedstock comes in and is processed in such a manner as to pull out the components, each of which may have commercial value. Take crude oil processing, as was mentioned before (Figure 1-4). In this case, the feedstock is crude oil pumped out of the ground. It consists of a number of different hydrocarbon combinations. Most of the individual components in the crude-oil stream have market value when separated from the other components. This separation occurs by heating up the crude to the point of vaporization and then cooling the vapor as it rises through a cooling tower.

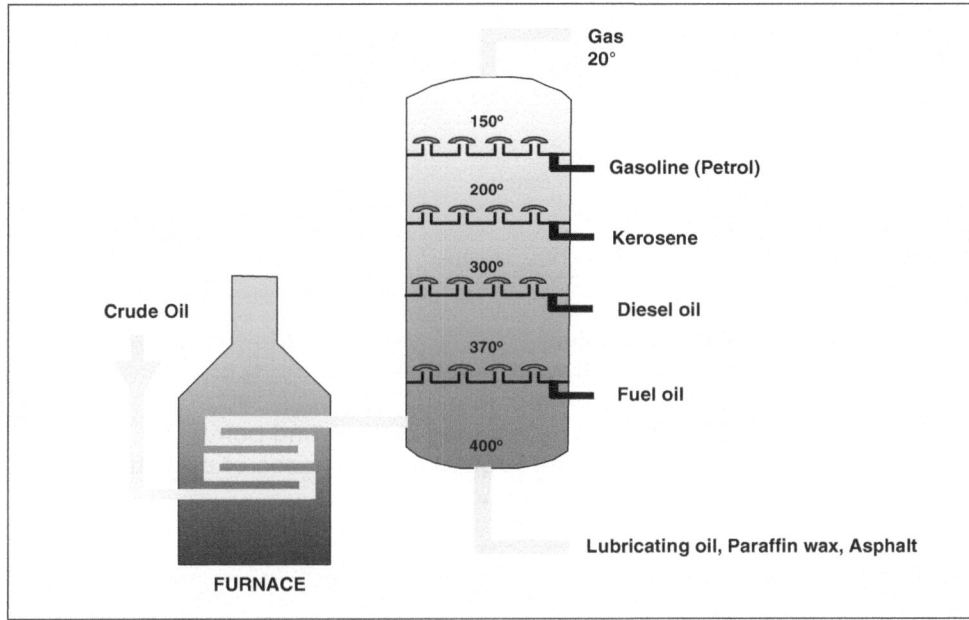

Figure 1-4 Continuous Process Example

Different components of the crude stream condense at different temperatures. As they condense they accumulate in a catch pan and are drawn off. In this case the salable products are fuel oil, diesel oil, kerosene and gasoline with additional byproducts that are converted into products such as lubricating oil, paraffin wax, and asphalt at later stages.

This disassembly process is typical of other continuous processes such as zinc processing, in which the feedstock is ore and the products might include zinc, other metals, and sulfuric acid. In paper mills, the feedstock is small particles of wood that are converted into liquid pulp, and the products are paper, bark, pulp, and turpentine. Many continuous operations involve multiple trains or lines that may interact with each other, which can increase the complexity of the operation.

Batch or Discontinuous Processes

A second type of manufacturing process is the batch or discontinuous process. As the name implies, manufacturers make products in batches or lots as compared to product being continuously produced. Unlike continuous processing, which often involves the disassembly of feedstock into base components, batch processing typically involves assembly-based processes using fluid and dry raw materials and the production of a single product at a time through the process equipment (Figure 1-5).

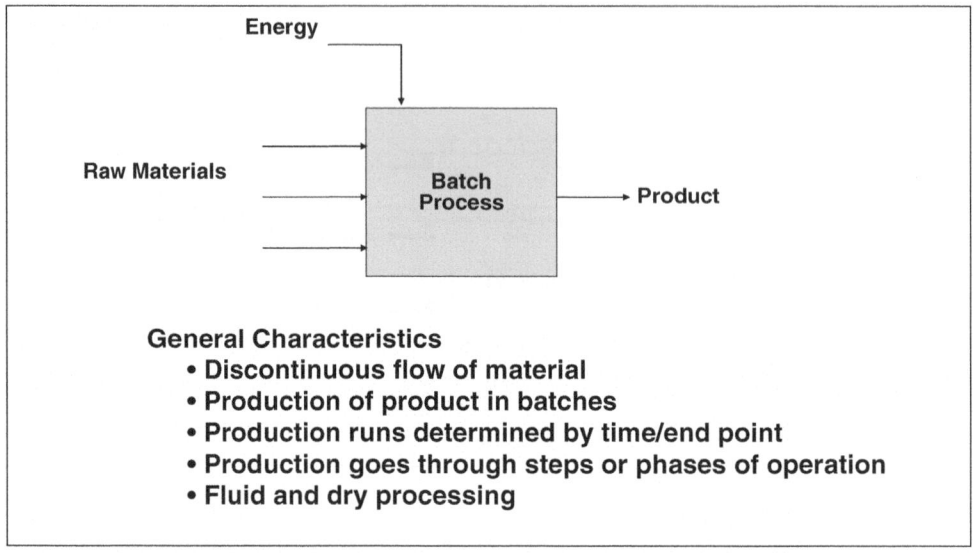

Figure 1-5 General Characterization of Batch (Discontinuous) Processes

Batch plants are those with a preponderance of batch processing and may include multiple process trains that can operate in parallel. A train is a collection of process equipment used to process a complete batch of product. Therefore, even though batch processes typically produce a single product at a time, a plant often simultaneously produces multiple products through different process trains.

Batch processes consist of a discontinuous flow of raw and processed materials. The raw materials are ingredients; each is typically introduced sequentially into the process in a prescribed order, and in prescribed amounts. This is the recipe. The order of processing is typically referred to as the phases of operation or the steps of the process. Sometimes a step is considered to be a segment of production within a phase. The ingredients come together to produce an expected quantity of finished product. With batch processes, a predetermined endpoint, usually defined by time or by the value of one or more process variables, determines the end of production.

Perhaps the simplest way to think about batch processes is to consider baking a birthday cake. In making the cake, we use a set amount of ingredients, mix them in a predefined sequence, charge them to a cake pan and put the pan in the oven until the cake is ready to help celebrate the 8th anniversary of our spouse's 39th birthday. The phases of this operation would be mixing, baking, and cooling. The endpoint of the baking phase is either determined by the time the cake is in the oven, or by an analytical endpoint measurement (sticking a toothpick into the cake) or both. We accomplish this batch process by following the recipe. An industrial cake baking plant would essentially do exactly the same things, in greater quantity, and would repeat the process to produce a larger quantity of cakes. In an industrial setting, candles are optional.

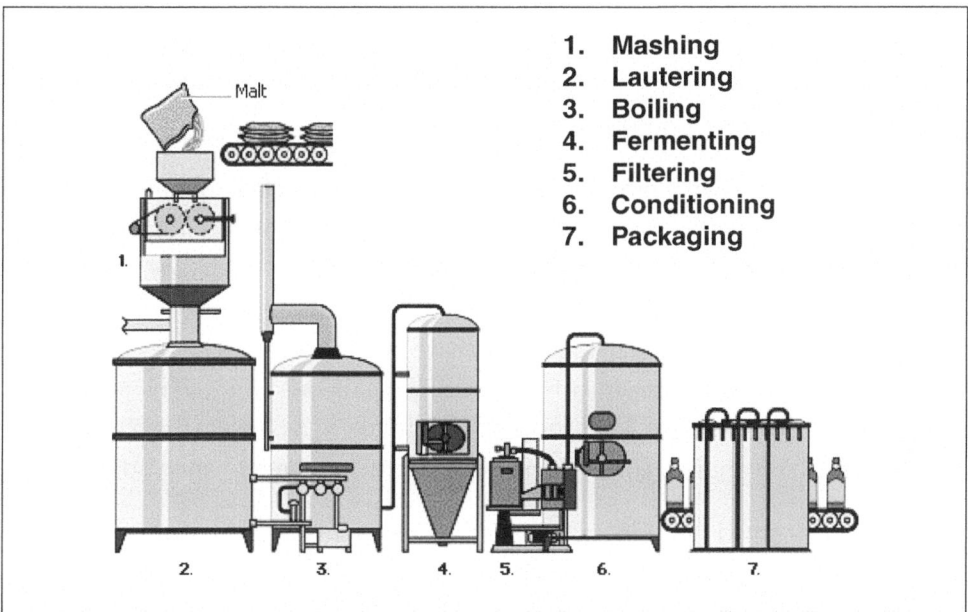

Figure 1-6 Batch Process Example

Beer brewing is a classic batch process that typically proceeds through seven phases of operation as the materials move from vessel to vessel in the brewery (Figure 1-6). The first phase is mashing, and it involves the mixing of milled grain and malt with water and heating the mixture to allow the enzymes in the malt to break down the starch in the grain into sugars. Phase two, lautering, involves the separation of the water and sugars from the spent grain. The third phase is the boiling of the extracts (called worts) to ensure that the mix is sterile, and the adding of hops to the boiling mixture to control flavor and aroma in the mixture.

The fourth phase is fermentation, which begins by adding yeast to the cooled wort, causing the sugars in the mixture to be converted into alcohol and carbon dioxide. The endpoint of the fermentation phase is determined by the time the mixture has been in the fermenting vessels. After fermenting comes the conditioning phase, in which the mixture is further culled, causing the yeast to settle and the proteins to coagulate, improving the smoothness and flavor of the beer. The beer is then filtered to remove impurities and finally packaged. This is a classic batch operation involving the sequence of a number of phases and the charging of different ingredients in predefined quantities.

Batch processing is much more flexible than continuous processing because a manufacturer can make a different product or product grade, with each batch made through the same equipment. In a brewery, a single type of beer is produced through the process equipment at one time. Most breweries produce multiple types of beer through the same process equipment by varying the recipes used.

The downside to batch processes is they tend not to be able to get the high levels of production that continuous processes can, simply because they do not produce continuously. Increasing the production rate of batch processes is typically limited by the size of the batches, the time it takes to complete a batch (the cycle time), and the time between batches. Batch operations can have different batches in the same batch train at the same time in order to increase the production rate. This can be done if the different phases use different plant vessels. For example, in a brewery there may be a batch of beer in the mashing phase while another batch is the fermenting phase.

It should be noted that a manufacturer that makes products in a continuous process could also make them in a batch process. The refining of crude oil is essentially a distilling process. Moonshiners have run batch stills for years, and batch stills could distill crude. Batch processing may be the better plant design if flexibility and agility are more important business issues than pure production. There is some thought in this day and age of custom manufacturing that batch production principles will again lead the way.

Discrete Manufacturing Processes

Discrete manufacturing is generally what people think about when they think about manufacturing. It involves the assembly of piece parts into products (Figure 1-7). Discrete manufacturing incorporates the staged assembly of products through a series of work cells. Each work cell has the assembly equipment necessary to complete one stage of the manufacturing process.

Discrete processes tend to be much more parts-oriented than the other two types of manufacturing and much less energy intensive. Discrete manufacturing operations offer the flexibility of batch processing and have some of the flow-through characteristics of continuous processing, but the flow is not fluid. Rather, it is a product being assembled. In a moving assembly-line plant, the product itself also moves through the various stages of production as it is coming together.

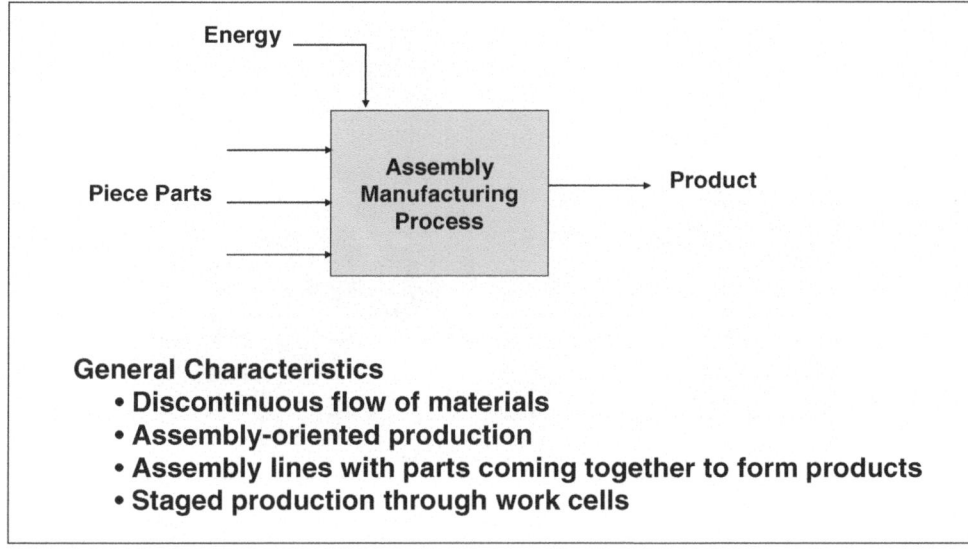

Figure 1-7 Generalized Characteristics of Discrete Processes

As with other types of production, one of the economic drivers of discrete manufacturing is to maximize the flow of product through the overall manufacturing operation in order to maximize production. In most cases, human operators train on how to operate in a single work cell and perform the same basic functions repetitively as each new, partially assembled product arrives. Unlike continuous and batch processes, discrete manufacturing processes are very visible to the worker involved in the operations of a work cell. Discrete manufacturing can be effectively implemented in a completely manual manner without the need for any automation or instrumentation. Automation and instrumentation come onto the scene to make discrete manufacturing operations more efficient and effective by improving speed, quality, and repeatability.

An automobile plant (Figure 1-8) is a good example of a discrete manufacturing process. Within a discrete manufacturing plant there may also be batch and continuous operations, such as wastewater management or even painting, but most of the operations are discrete. Automobiles are pieced together on assembly lines characterized with work cells, each completing one stage of the manufacturing process. These stages might include welding the frame together, installing the wheels, and installing the motor on the frame. Manufacturing engineers focus on making sure the parts are in the right place when required, the assembling product is flowing as efficiently as possible, and the work done in each cell is done to the desired quality.

Figure 1-8 Discrete Process Example

In discrete manufacturing processes, manufacturers define quality in terms of defects per millions of parts made. The concept of "defects" does not translate directly in batch and continuous processing since they are fluid processes rather than parts-based processes. There are approaches to discrete manufacturing other than assembly lines, such as fabrication shops, but from an industrial automation perspective, assembly-based discrete processes are the most interesting.

Manufacturing Processes and Industries

As mentioned, you can often find the three types of manufacturing processes in any industrial plant. Figure 1-9 shows a continuum of industries positioned according to the level of automation commonly employed and the type of manufacturing implemented. Plants that tend to have a predominance of continuous processing are in the refining, bulk chemical, gas, power, paper, and mineral processing industries. These industries have been the ones for which the economic proposition has clearly been driven by production volume. Notice that these industries are also in a sector labeled as scientifically-oriented manufacturing. This means that significant scientific analysis went into the design and operation of these processes, and they are well understood in terms of the science behind the production. Most continuous process plants have a high degree of industrial automation.

Toward the middle of the chart are the craft-oriented industries. There are a number of different industries included in this group, such as fine chemicals, bulk pharmaceuticals, beverages, and biotechnology. These industries have a mix of

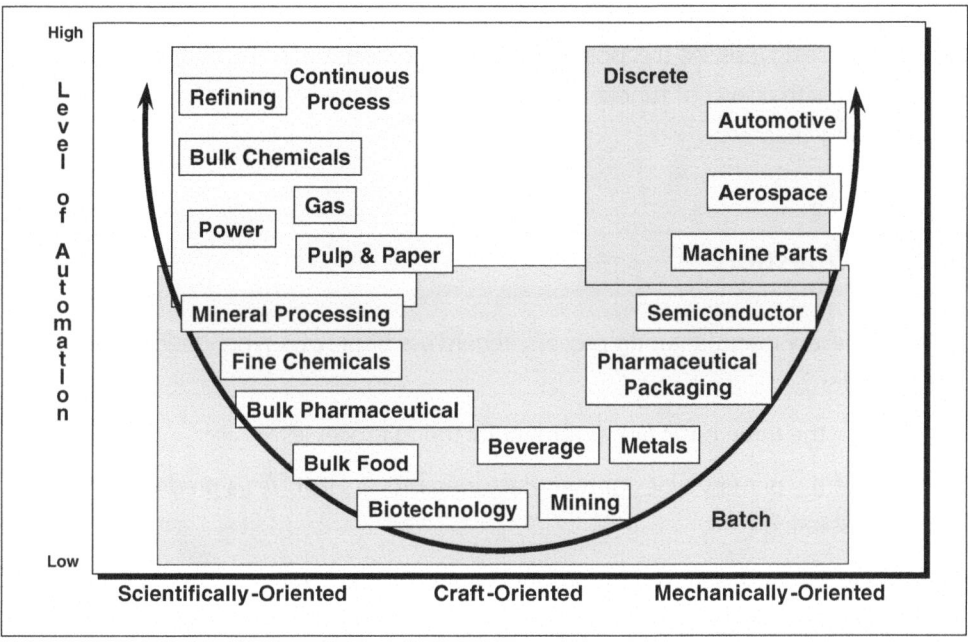

Figure 1-9 A continuum of industries positioned according to the level of automation

processing types but batch often dominates. They are craft-oriented industries because, for the most part, they have not had the level of scientific analysis done with respect to the manufacture of the products, resulting in the manufacturing being more of a craft than a science.

The final category, at the top right of this chart, includes the industries that are more discrete process-oriented. These tend to focus on piece parts being mechanically assembled into products and are categorized as mechanically-oriented industries. The scientifically-oriented industries and the mechanically-oriented industries tend to have greater levels of automation employed than do the craft-oriented industries. This may be because it is easier to automate manufacturing processes that have been scientifically or mechanically well defined compared to the craft-oriented industries, but it is more likely due to the fact that the economic value proposition for automation in continuous and discrete plants has tended to be greater than that for batch plants.

Understanding the basic manufacturing processes and how they come together in manufacturing plants is important to the study of industrial automation. Much of the terminology used in industrial automation comes from basic manufacturing terminology, and different automation technologies have been designed for the different process types. At this point, the material presented in this chapter should provide you with a level of functional understanding necessary to move forward to the following chapters.

Review Questions

1. What is every manufacturing process, regardless of process type, designed to accomplish?

2. What are the three basic types of manufacturing processes?

3. Which of the primary types of manufacturing process involves a predominance of fluid processing?

4. Which of the primary types of manufacturing process involves the assembly of piece parts?

5. What is a manufacturing train?

6. Which of the three basic manufacturing processes is designed to maximize production volume?

CHAPTER 2

Control of Discrete Processes

The scene plays over time and time again. A shell of a car moves down the assembly line until it hits a familiar spot. The car slows, and the robot swings over with a tire, ready to place it in the proper position. After the tire is on, the car moves on down the line for the next part installation, just another step in the birthing process of a car. While on the face of it, the act of piecing an auto together seems complicated, it really isn't. As we have seen, discrete manufacturing is typically assembly line oriented, involving the progressive assembly of finished products via a number of work cells, each providing a predefined set of mechanical operations. These mechanical operations frequently involve the use of machines of different types, designed for the functions to be completed in each work cell. Typical machines might include lathes, welding machines, cutting machines, power wrenches, robots or lifts. Controlling discrete manufacturing processes essentially constitutes the automatic operation of these machines. Automation of these operations is largely a matter of a series of carefully timed on-off steps. Initially assembly-line workers executed these steps; however, through a series of technological innovations since the 1960s, it all became more automated.

Although there have been a number of very clever electronic and electromechanical devices invented for the automatic control of discrete manufacturing processes, the most prevalent are the automatic switch and interconnected sequences of automatic switches. Although devices such as automatic timers also see use in discrete process control, their operation and construction are fairly straightforward; therefore, since automatic switching is the heart of discrete process control, that is what we will address.

The first automatic switch to be extensively used in discrete process control was the electromechanical relay (Figure 2-1). A number of electrical suppliers, such as Allen Bradley, Siemens, and Square D, manufactured these devices for use in discrete manufacturing control as well as for other areas requiring high-speed switching. Although the diagram in Figure 2-1 is a simplified view of an electromechanical relay, it is useful for understanding the basic principles of operation of these devices.

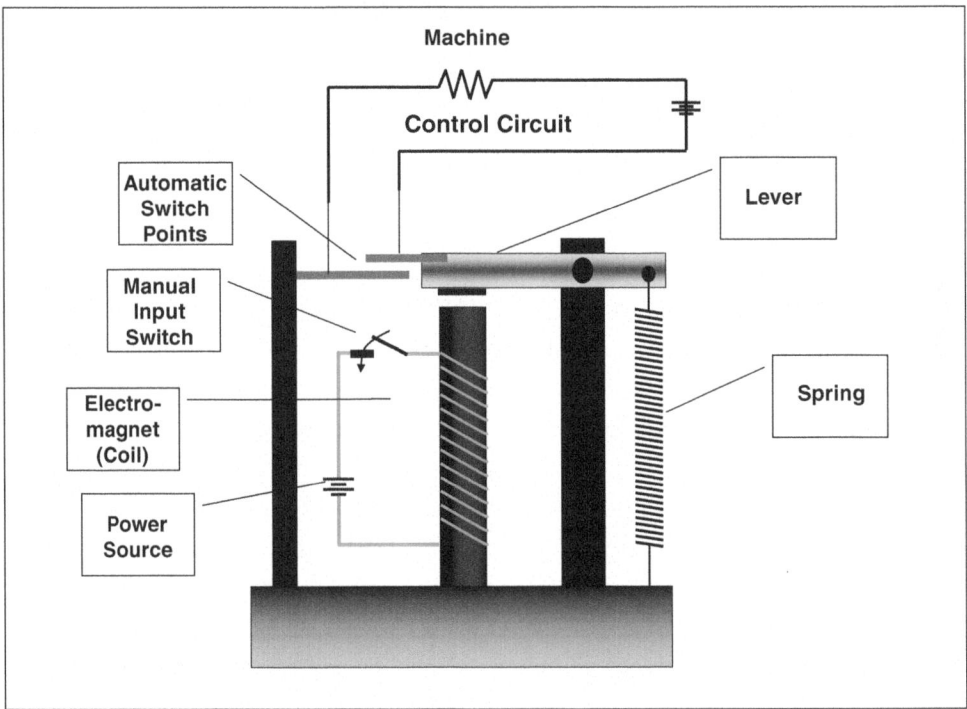

Figure 2-1 Operating Discrete Processes — The Electromechanical Relay

The fundamental driving element of these relays is an electromagnet. An electromagnet is formed by wrapping an iron-based bar with electrical wire. With electromechanical relays the electromagnet is sometimes referred to as the coil. When the input switch is closed, electricity moves through the wire, and the iron bar becomes magnetized.

The automatic switch mechanism of the electromechanical relay is formed by the construction of a lever attached to a fixed post. The lever has a piece of iron attached to it that is attracted by the electromagnet when it is in the magnetized state. The back side of the lever has a spring attached to it that normally pulls the back side of the lever down, forcing the switch side of the lever into the "up" (off) position. When the electromagnet is in the magnetized state, the magnetic force pulls on the lever, overcoming the pull of the spring and causing the lever to move down toward the electromagnet.

With one contact on a separate fixed post and the other on the lever, the automatic switch opens and closes according to the magnetized state of the electromagnet. This results in an automatic switch. This description is fairly accurate, but it is a bit of a simplification in that a single electromagnet may provide the driving mechanism for multiple sets of contacts. In such a design, when a single electromagnet is magnetized, a number of different on-off actions can occur simultaneously.

It does not take much imagination to consider how a number of electro-mechanical relays could be connected together to perform a number of actions at the same time. As we have seen, the driving element of an electromechanical relay is an electromagnet powered by a circuit engaged through an input switch. Therefore, if the input switch were another electromechanical relay, activating that relay would provide a chained response. In other words, these relays can be strung together in very clever ways to perform a number of different binary (on-off) operations. To make this even more interesting, if timing mechanisms, such as the spring-type clocks often used in the kitchen as a cooking timer, are inserted within the relay systems at various points, the actions driven by a single push of a button might drive a large sequence of events that are very carefully timed.

One aspect of electromechanical relays that is very useful in developing these relay systems is that certain actions may require an electrical switch to be turned on while other actions may require an electrical switch to be turned off. For example, suppose an operator wants to start a welding machine by pushing a start button on the side of the machine. Suppose there are warning lights over the work cell, one red to alert others when a welding operation is taking place and the other green when the welding machine is off. When the operator pushes the start button, it should power an electromagnet that turns on the red light and turns off the green light at the same time. Two basic electromechanical relay options have been designed to accomplish this, as is shown in Figure 2-2. The two options are referred to as "normally open relays (or contacts)" and "normally closed relays (or contacts)." With a normally open relay, the contacts are attached to the lever and post so that when the electromagnet is not magnetized, the switch is in the open position. The opposite is true for the normally closed relay.

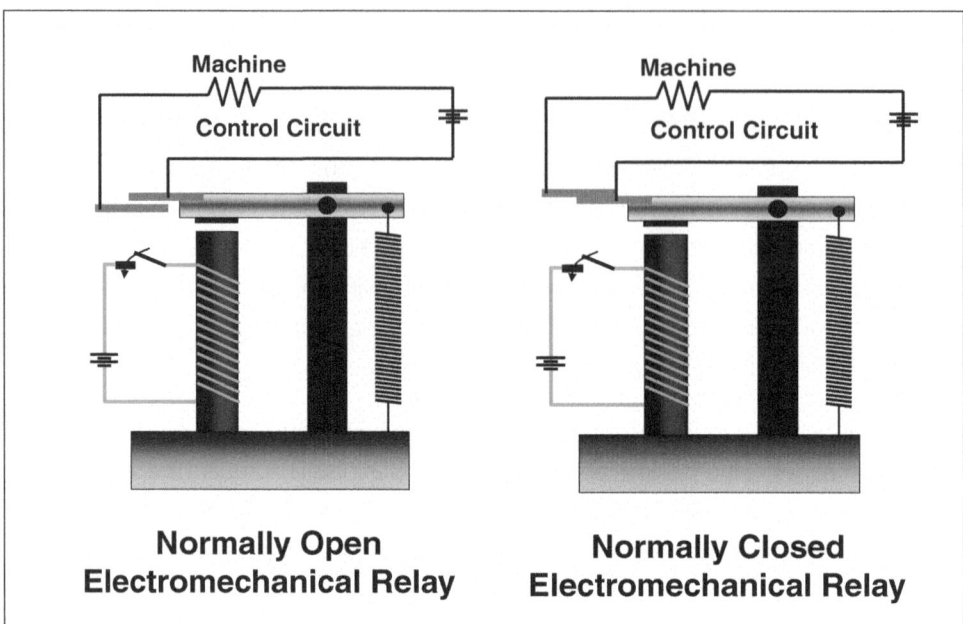

Figure 2-2 Types of Electromechanical Relays

The availability of "normally open" and "normally closed" relays allows electricians to design very complex logical sequences into their control systems by connecting both types of relays and timing devices together. Collections of relays that combine to perform automatic logical operations like that described provided the basis for the first automatic control systems in discrete manufacturing operations. Electricians typically assembled the relays on relay panels in a manner that would accomplish the desired automation steps. The electricians called the assemblies of relays on the panels "relay ladders," and often became proficient at developing very complex logical schemes for the automation of work cells. They would work out the circuitry on diagrams using symbols for the switches and coils and would refer to these diagrams as "ladder diagrams." The diagrams often provided documentation for the electromechanical circuits the electricians installed.

The electromechanical relay control systems worked very well, but they had a few shortcomings. As with any mechanical element, after prolonged use they tended to wear out or break. In complex systems the repair of the relays could be very difficult. As engineers and electricians became more proficient at designing and implementing these systems, the systems tended to increase in complexity and often required considerable room to house. Also, these systems could get quite expensive as more and more relays were required to build complex logic schemes. Electrical suppliers were looking for ways to address these shortcomings as the price of digital computer technology started to come down to the point at which it was starting to become popular for many applications. The basic active components of digital computers are binary switches and circuits. These computers presented a natural solution for a new wave of discrete manufacturing control systems.

In the 1960s, however, digital computer technology was not only expensive, it also had a poor reputation for reliability. Reliability is a critical issue in manufacturing. Although it seemed like a good idea to apply digital computer technology to replace electromechanical relay systems, manufacturers were reluctant to apply any unreliable technology in their manufacturing operations. To solve this problem, the manufacturers of computer-based logic controllers decided to disguise the fact that they were utilizing computer technology by calling these devices "programmable controllers" or "programmable logic controllers." Richard Morley of Modicon Corporation in Massachusetts is credited with the invention of the programmable controller in 1968. The first programmable controllers were advertised as "solid state" devices, which has a nice stable sound to it. Referring to them as computers was avoided at all cost. Many of the companies that marketed the first programmable controllers were the same electrical suppliers who had been marketing electromechanical relays and other electrical equipment.

Anybody who has studied binary numbering systems and logic, which underpin the workings of digital computers, will easily understand why such a device would be ideal for complex binary logic operations. The digital computer is a complex binary logic machine. The problem in the 1960s was that computer programmers were scarce and very expensive. The challenge was in developing the computer so electricians who already had a good working knowledge of electromechanical relay logic systems and their design would be able to set up the programmable controllers. The solution was simple, but ingenious. A computer programming language was developed to emulate the relay ladder diagrams the electricians had been using to set up the electromechanical relay systems (Figure 2-3).

The initial language of programmable controllers was called ladder logic. Electricians would develop their relay ladders as they always did, then go to a programming panel for the programmable controller and draw essentially the same diagram on the programming panel. The program would then be compiled, or downloaded into the programmable controller. Once the program was in the programmable controller the programming panel could be unplugged and the programmable controller would just sit there and operate. Since there are no moving parts in programmable controllers, they actually proved to be much more reliable than the older electromechanical relay systems. Also, for complex logic systems they tended to take up much less room than electromechanical relays did. As the technology has developed, they have become much smaller and much less expensive than they initially were.

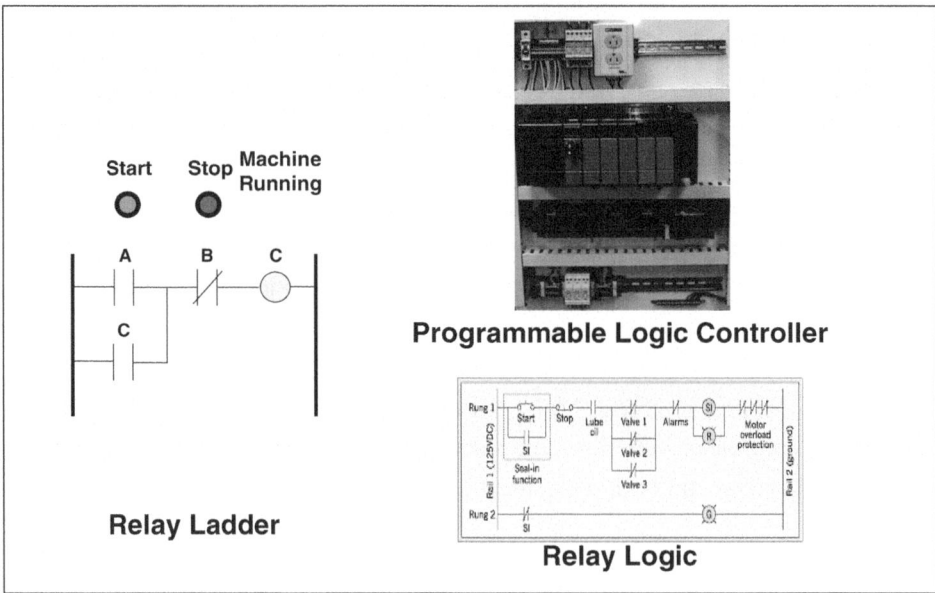

Figure 2-3 Programmable Logic Controllers

Figure 2-4 shows some older photographs of manual assembly lines above some photographs of more automated assembly lines. These assembly lines have been automated by programmable controllers that are not even visible in the picture. Notice how few people are required in the automated line as compared to the manual line. Although relay logic systems led to a limited level of headcount reductions, PLC systems allowed much greater levels of reduction. The cost savings in headcount alone would often provide justification for the installation of either of these automated systems. The increases in production rate and the reliable repeatability of repetitive operations made the application of this technology one of the best economic investments many discrete manufacturers had at their disposal.

Most automated discrete manufacturing operations today still employ programmable controllers as the primary automation device. In many cases these

Figure 2-4 Examples of Discrete Process Control

devices still sit unattended, performing their job through pushbutton interfaces. As shall be discussed a bit further on, programmable controllers are now being integrated with personal computer technology to form computer-based control systems for discrete manufacturing.

Review Questions

1. What is the primary control device in an electromechanical discrete control system?

2. Why are normally open and normally closed relays both required to set up discrete control strategies?

3. When a number of electromechanical relays are wired together into a control strategy, what is the resulting system called?

4. What is the name of the computer that performs the same control functions as a bank of electromechanical relays?

5. What is the computer language that is programmed in a similar manner to electromechanical relays?

CHAPTER 3

Control of Continuous Processes: Stay In the Loop

When it all comes down to it, the most important thing that determines the optimum operation of a continuous process plant is the setting of set points and maintaining the controlled variables at the set points. Set points and control are the keys. They determine the values of the flows, levels, temperatures, speeds, and other variables across the plant.

But before we get too far ahead of ourselves, let's take it from the top. While the control of discrete manufacturing processes tends to involve logic-based operations due to its on-off nature, the control of continuous processes tends to be scientifically based. Continuous manufacturing processes involve some level of chemistry or physics, making control of these processes much more mathematical than logical.

Since most continuous process and batch manufacturing occurs within pipes or vessels, it tends to be invisible to the process operator. And since all operators are not capable of employing Superman's x-ray vision, they need some degree of instrumentation to make visible certain physical and chemical aspects of the processes. Initially, companies that became involved with the automation of continuous processes were companies that made instruments to measure flows, levels, temperatures and pressures within the piping and vessels and to make these measurements visible to the operators. These companies included Honeywell, Foxboro, Taylor Instruments, Rosemont, Fisher Controls, Masoneilan, and Bailey, among many others.

Basic Feedback Control

The basic control entity in continuous processing is the feedback control loop. At its most fundamental, a feedback control loop is the set of functions required to measure and maintain a single process variable (flow, level, temperature, pressure, etc.) at a specific desired value, called the set point. Way back when, the earliest control loops were manually operated (Figure 3-1). The measurement was made by a process instrument and displayed on a dial attached to the instrument. The process operator standing at a manual control valve looked at the dial to determine the value of the process measurement (measured variable), compared the current value to a predetermined desired value (the set point), and turned the valve in the direction that would make the measured value approach the set point. Once the valve was adjusted, the process changed accordingly, reflecting the new valve setting, and the value of the process measurement likewise changed. The measurement change was then reflected on the dial. When the operator got feedback on the new value of the process variable by looking at the dial, the operator adjusted the valve again, if necessary, until the set point and the process measurement came into alignment. It is easy to understand why this system is referred to as a "feedback control loop," since effective control involves looping through the various elements involved and adjusting the process based on feedback as to the impact of previous adjustments.

A simple feedback control loop consists of just a few basic components: the process itself, one or more measuring devices, some means of adjusting or changing the process, and a controller. It is helpful to understand the basic components of the control loop and establish a vocabulary that will enable us to effectively build from a simple control loop to higher levels of process control.

The basic components of an automated process control loop are shown in Figure 3-2. The first component is the process itself. The process provides the physical or chemical transformation required to make a product or products. This transformation could be as simple as the heating of a liquid, or it could involve an aggressive exothermic (heat-producing) reaction between chemicals. In any case, the process exists to make the product.

It may seem as though once a valve is positioned correctly there would be no need to adjust anything, and the process would just keep operating as desired. Unfortunately, things change around the process that may cause the measurement to change even though the valve has never moved. These changing conditions in and around the process that impact the measurement are called process loads. Process loads are physical, mechanical or chemical conditions that are typically not controlled, but that can impact the value of the process measurement. For example, a change in temperature in the room in which the process is operating can cause the temperature of a liquid in the process to change. The temperature in the room is a load on the process.

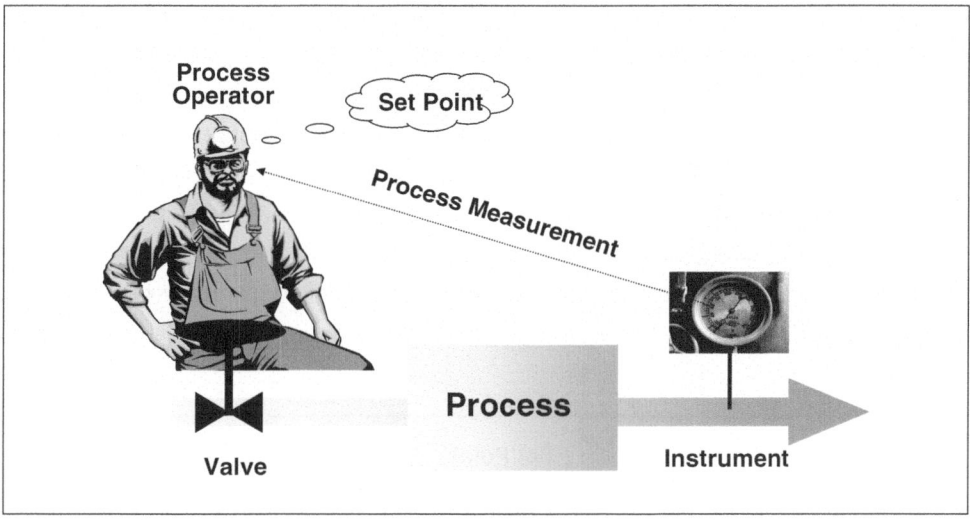

Figure 3-1 Manual Feedback Control Loop

In an automated control loop, as in a manual one, the measured variable is the one being controlled. In order to control it, its current value has to be measured. This is accomplished by the second basic component, the measurement instrument. The instrument provides a signal to a process controller, communicating the current value of the measured variable. The controller compares the current value of the measured variable to the desired value—the set point—and determines if the process needs to be manipulated to bring the measured value into alignment with the set point. If so, the controller sends a signal to an actuating device, typically one operating a valve, which drives the change required. Through the valve, the actuating device manipulates a variable in the process, and the manipulated variable changes the process in such a manner as to cause the value of the measured variable to change.

There are two primary dynamic characteristics of any continuous process that impact the control of the process: dead time and lag. Dead time is the amount of time it takes for the process to react to a change, typically a change in a manipulated variable. For example, a process controller may make a request to a valve to close by 5%. The flow along the length of pipe between the valve and the process may consume ten seconds, so the process is not impacted until ten seconds after the request is made. This delay would introduce ten seconds of dead time into the process loop.

Lag is a measure of how quickly a process responds to a change and is normally related to the volume (capacity) of material or energy in the process. For example, a large tank containing thousands of gallons of a chemical mixture would have a large capacity. If a small pipe at the top of the tank is adding a component chemical in order to change the mix in the tank, opening the valve in that pipe would cause more of the additive to be charged into the tank, but there would be a lag in the time it takes to discern a change in the chemical mixture because the capacity of the tank is so large compared to the capacity of the inlet pipe. In a simple sense, dead time and lag impact the controllability of the process in opposite ways. The more dead time there is in the process, the harder it is to control, while more lag makes it easier to control.

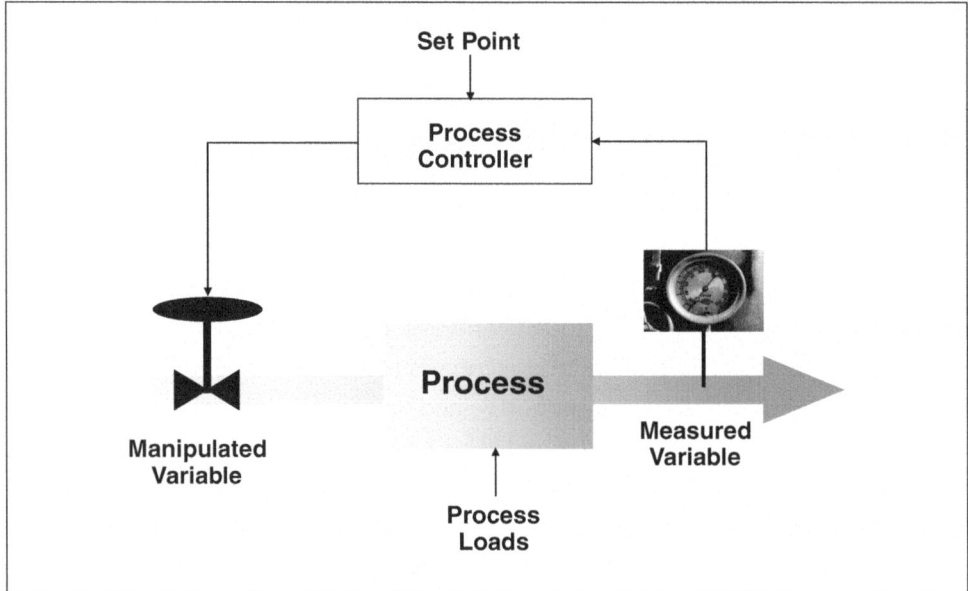

Figure 3-2 Basic Feedback Control Loop

Since continuous process manufacturing plants operate in an ongoing manner and require the holding of multiple process variables to specific values to make products effectively and efficiently, the control of most continuous process plants can be thought of as the coordinated control of a number of control loops. Controlling continuous process plants has long been approached by controlling one loop at a time until the entire plant is under control. The challenge is to make sure each loop is well controlled and the set points of all the loops are set to values that optimize the operation of the plant as a whole. Any discussion of the accepted approach to the control of continuous process plants begins with the basic components making up the various elements of a control loop.

Process Measurement and Instrumentation

A measurement cannot be controlled if it is not measured. Therefore it makes some sense to start with the measurement aspect of the variables to be controlled. Process measurement is done by process instrumentation. The four primary measurement types in process plants are flow, level, pressure, and temperature. Over the past thirty years, however, a number of additional measurements have become available through the development of more sophisticated measuring technologies using several different techniques ranging from analytical chemistry to nuclear magnetic resonance. The job of the instrumentation is to provide measurements of the key process variables and to transmit a signal indicating the value of the process measurement to process controllers. In the previously discussed manual control loop, the measurement is made and displayed on a dial. The transmission of the measurement signal is simply accomplished by the operator looking at the dial. Figure 3-3 shows some typical measurement instruments.

One key challenge in measuring process variables is the processes are operating at the time the measurements are made, which can cause noise in the signal. Noise is distortion in a signal due to unintended conditions. For example, trying to measure the level in a tank while liquid is flowing into the tank can be difficult because of turbulence. Part of the control challenge is to control (or compensate for) noise in the measurement. But at the end of the day, the job of each of these instruments is simply to measure some process variable, or set of process variables, to a desired accuracy and send a signal representing the value of the measurement to a controller.

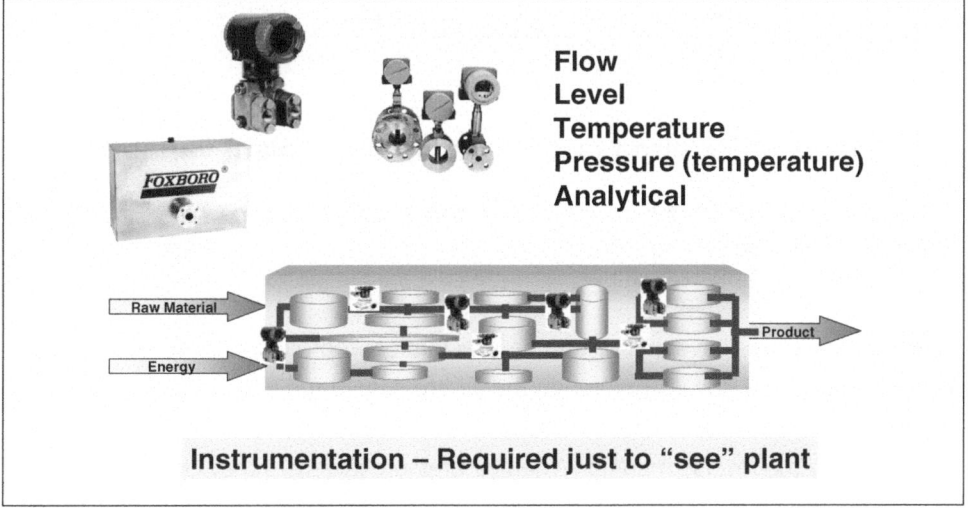

Figure 3-3 Process Instrumentation

Devices that Adjust the Process — Valves

The third set of primary components of a process control loop is the adjusting devices that alter the manipulated variables in the processes. For the most part these are valves (Figure 3-4), but they can also include such devices as air dampers that control air flow and motor drives that control the speed of motors. A fairly large variety of control valve mechanisms and suppliers exist in the marketplace, but the basic operation of these valves is fairly consistent across this broad spectrum. The mechanism on the valve that actually drives its position is called an actuator. The valve is in the flow stream to be controlled, and the actuator receives a signal from a process controller that indicates in which direction (open or closed) and by how much the valve should change from its current position. The majority of valves in use in process manufacturing plants today are air-powered, or pneumatic. The pressure of the air signal they receive determines the desired position of the valve.

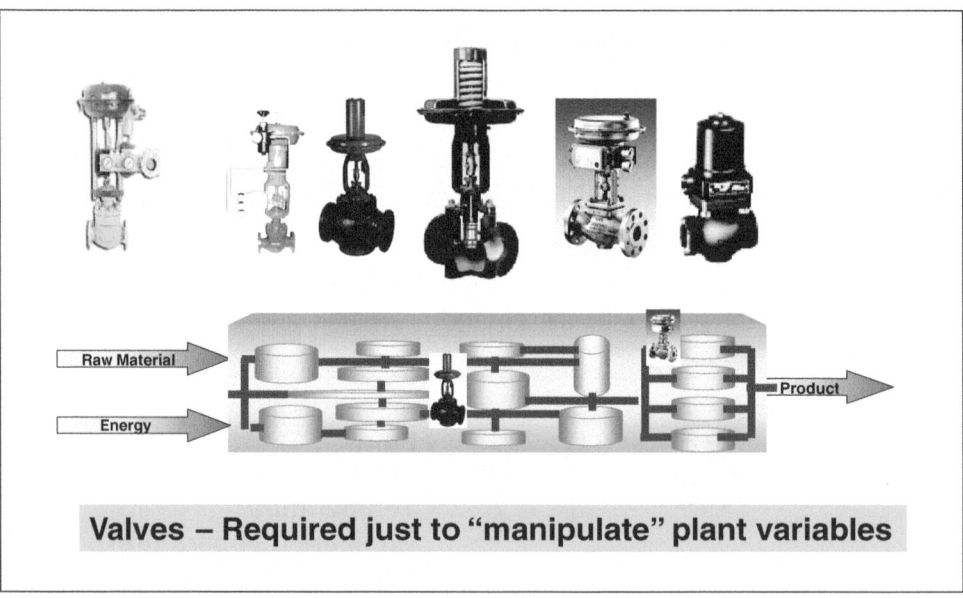

Figure 3-4 Process Adjusting Devices — Valves

Pneumatic valves have one of two basic actuation approaches: air-to-open or air-to-close. Air-to-open valves open when the air pressure in the input signal increases and close when the air pressure decreases. Air-to-close valves, on the other hand, react in exactly the reverse manner. When air pressure increases they close and when it decreases they open.

Valve movement is accomplished through the use of an actuator that consists of a spring mechanism behind a gasket. When air is introduced into the actuator, the gasket pushes on the spring mechanism to cause it to compress, moving the valve. The mechanism is in one orientation for air-to-open and in the reverse orientation for air-to-close. The reason for these two approaches to actuator design is to support safe operation of the process. Since the valves require a pressurized air source to operate, it all depends on which position the valves should be in to ensure a safe state in the case of air pressure loss to the valves. If a valve enables the flow of a volatile chemical into a vessel, the most appropriate action in the case of a pressure loss would typically be to close the valve so the volatile chemical ceases flowing. To accomplish such an action, an air-to-open valve would be selected. This first level of process safety falls right into the engineering design of process plants through the appropriate selection of valves and actuators.

The Process

The final component of a basic process control loop is the controller itself. In the manual control loop, the controller was the person looking at the dial on the measurement instrument and adjusting the valve to try to align the measurement with the set point of the loop. Automatic control loops required replacing the person as the controller with an automatic control mechanism. These automatic controller mechanisms can be as simple as the mechanical link devices that control the flow and level in a toilet. Automatic controllers have also been produced using pneumatic (air), analog electronic and digital electronic mechanisms. In any case, these devices receive an input signal from the process measurement device, which represents the measurement of the process variable of interest. It also receives an input signal, often input by a process operator, which indicates the desired value (set point) of the measurement. The measurement value is compared to the set point value, and a signal to the valve is determined and transmitted by the controller. If the measurement and set point are not aligned, the output signal will cause the valve to move in the direction (open or closed) that will cause the measured variable to move toward the set point. It is somewhat like a leveling device, where the set point is "perfectly level." To reach the set point, a controller will continue to send out signals to the valve until the measured variable reaches the "level" or set point.

The transmission signals (pneumatic, electronic analog, and digital electronic) used in many automatic control systems allow the signal to be transmitted over distance, enabling multiple controllers at a single control station to be managed by a single operator (Figure 3-5). Automatic control systems configured in this manner are preferable to manual control for two reasons. First, automatic controllers operate continually without loss of efficiency due to tiredness or boredom, as is the case with humans. This results in better control. Second, not as many operators are required. Often the cost savings due to headcount reductions alone justify the expense of automatic control.

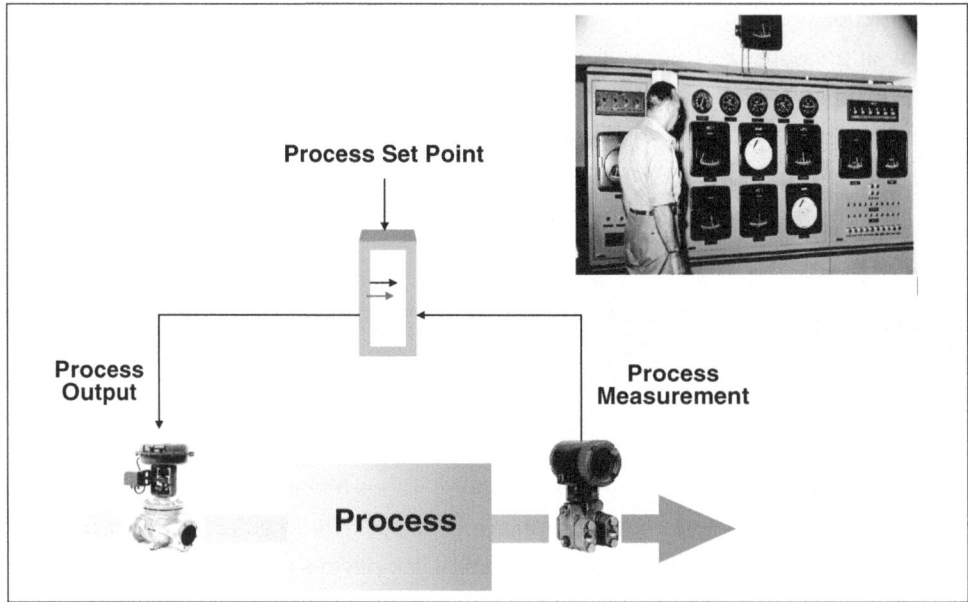

Figure 3-5 Automated Process Control Loop

   Operators are responsible for monitoring the overall operation of the controllers in their area, setting the set points of the controllers and taking care of abnormal situations. Controllers can also be switched into manual operation, enabling operators to control the valves by directly driving the output signal to the valve from the controller faceplate. It is not unusual for operators to bypass much of the automatic control because they prefer running the plant in manual. This has been a problem ever since the introduction of automatic control systems. Plants seldom operate as efficiently in manual as they do in automatic. Trusting the technology to run the plant has been an issue for years. While the technology is out there to control a process, as mentioned, operators will often try to run the process in manual. As automation continues to infiltrate the plant, and there are fewer people running plants, operators will be forced to run in automatic control.

Feedback Control Example: Heat Exchanger

To return to our discussion of the operation of a control loop, a heat exchanger, which is a fairly common and simple piece of equipment, will serve as an example. A heat exchanger is a vessel designed to heat the process fluid (in this case, water) flowing through it. This is accomplished by pumping the water through a pipe that passes through a heated vessel (Figure 3-6). Steam is introduced into the vessel to provide the heat. The objective of this process is to heat the water passing through the heat exchanger to a specific temperature. The measured variable is the water temperature. A signal representing the value of the water temperature is transmitted to the controller. The controller compares the actual water temperature to the desired temperature (set point) and sends a signal to a valve. The valve regulates the manipulated variable—the steam flow into the heat exchanger. If the actual water temperature is lower than the set point, the signal from the controller opens the steam valve, introducing more steam into the exchanger, which causes the water to heat up. If the actual water temperature is greater than the set point, the controller closes the valve to reduce the steam flow into the heat exchanger, which reduces the heat input and allows the water to cool.

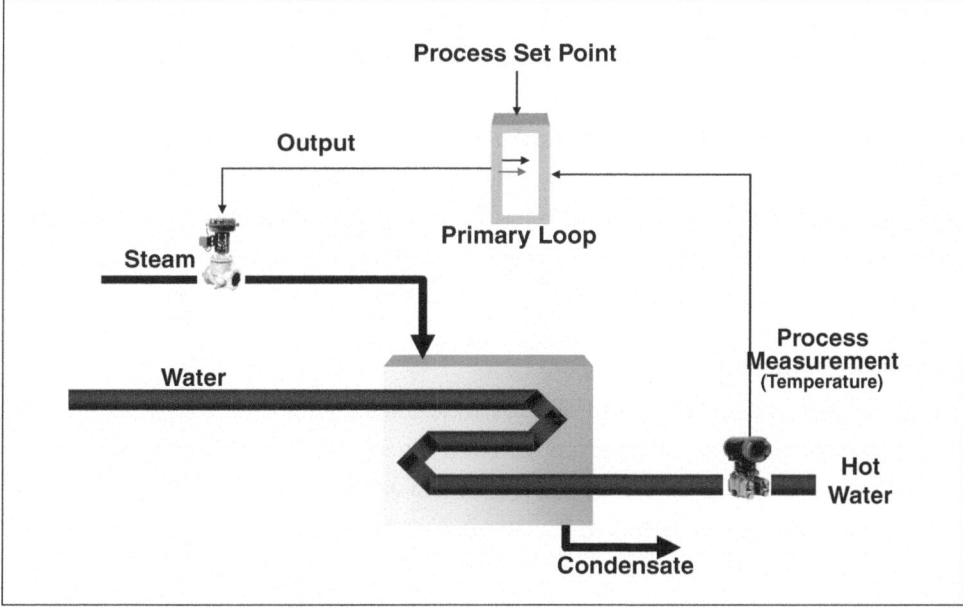

Figure 3-6 Example — Heat Exchanger

Once a control loop is set up and operating in a manner at which the set point and measurement align, it may seem as though no further adjustments would be needed. Unfortunately, control loops work within an overall plant in which many changes are taking place every minute. These changes may introduce a load change to the process being controlled. Load changes often cause the measurement value to change, moving it away from the set point. With the heat exchanger, a good example of a load change is an increased water-flow rate. With more water flowing through the heat exchanger, the output temperature of the water would drop, causing an error between measurement and set point and requiring a control action to adjust the steam valve. There are a number of different loads associated with a heat exchanger, such as water flow, steam pressure, ambient air temperature, and condensate buildup to name just a few. That all means there is a technological form of checks and balances going on at all times during a process run.

One of the issues the controller must be able to deal with when the measured water temperature is not at the set point is how far to open or close the valve. The amount of response the controller makes to an error between the measured value and the set point is the "control response." The control response depends upon the physical, thermal, chemical, etc., characteristics of the process being controlled. In the heat exchanger example, if the heat exchanger vessel is large and the steam flow is relatively small, it may take a few minutes for an increase in steam flow to have the desired effect on the temperature of the water. If, on the other hand, the heat exchanger vessel is small and the steam flow is large, a small adjustment to the steam valve would cause the water to heat quickly. In the first case a large adjustment would have to be made to the valve to realize the same response as would be accomplished by a much smaller adjustment to the steam flow in the second case. Setting up the controller to have the correct control response is called "tuning the loop."

Controller Tuning

Today, most feedback process controllers provide three coordinated responses to correct an error between the measurement and the set point: proportional, integral and derivative. The proportional response is one in which the controller adjusts the valve by an amount proportional to the amount of difference (the error) between the measurement and the set point. Proportional action is a commonsense response mechanism for control and would be the only feedback control mechanism required, except that as process loads change, a simple proportional adjustment results in an offset between measurement and set point. The offset is a difference between the measurement and the set point once the control action has had its effect.

Integral control action was added to correct the offset resulting from proportional action. Although it's a simplification, integral can be thought of as a slow adjustment back to the set point once proportional action has had its effect. Integral action corrects the offset, but takes a relatively long time to get the measurement back to the set point. In some cases the slowness of the integral response is undesirable. The derivative response provides a "kick" that was developed to speed things up a bit. Together they are referred to as PID control.

Perhaps the easiest way to explain PID control action is to use an everyday example, driving an automobile. Driving an automobile can be thought of as an application of manual feedback control. The driver is the controller. The driver of a car moving down a road controls velocity through the gas pedal. As long as the road is straight and level, the driver can maintain the speed by keeping the gas pedal at the same setting. If the car encounters a hill, as the car starts to ascend it will also start to slow down. The driver, sensing this slowing, will step on the gas pedal a proportional amount (proportional response) to the sensed slowing of the car. Typically this first response is not exactly correct to bring the car back to the desired speed so the driver slowly adjusts (integral response) the pedal to work the speed back to where the driver wants it to be. If this is taking too long, the driver might over-adjust the gas pedal to provide a kick (derivative response) to increase the velocity.

Tuning control loops is very important to the effective operation of a process. A feedback controller may be capable of a combination of proportional, integral and derivative responses, but it may not use all three. Each of the control responses used needs to be tuned in order for the control mechanism to provide the optimal feedback response. A control engineer typically would tune the proportional response first, followed by the integral response and then the derivative response if all three are used. Clearly the responses impact each other and each needs to be tuned according to the process and the tuning of the other responses. Experienced control engineers can often determine initial tuning settings that provide close to an optimal response. The generally accepted response to a well-tuned controller is a diminishing oscillation back to the desired set point (Figure 3-7).

Basic process control is implemented in process plants by controlling each process control loop and keeping the loops tuned to provide optimal feedback response. Software has been developed for the initial tuning of loops and for keeping them in tune by evaluating them over time and making the appropriate adjustments. This software is important because production processes change over time. Equipment wears out, raw material compositions may change, instrumentation may be updated, and many other things typically happen over time that change the dynamics of the process. As these dynamics change, the tuning of the loops needs to be adjusted to match.

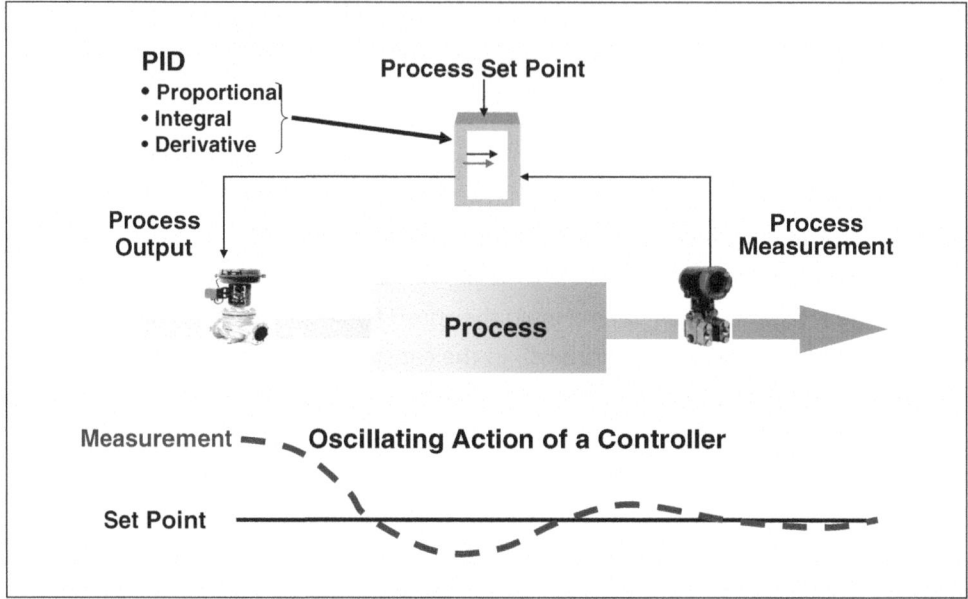

Figure 3-7 Control Loop Tuning

If the control loops are in place and tuned, the primary activity that determines the optimal operation of a process plant is the setting of the set points. As mentioned at the beginning, the set points determine the values of the flows, levels, temperatures, speeds, and other variables across the plant. If the set points are not set correctly, even if the controllers are correctly implemented and tuned, the plant is not operating in the best possible manner; and if the loop is not in control, it really doesn't matter what the set points are.

We have intentionally over-simplified the science behind the control of continuous processes in an attempt to make the overall subject matter more understandable for any person looking into this topic for the first time. For continuous control strategies to be effective you need a very high level of engineering expertise in the areas of controller tuning, control loop selection and design, operations performance across multiple control loops, interaction and decoupling between control loops, field device selection, specification and installation of instrumentation and valves, and plant-wide control operation, as well as many other important topics. As with any sophisticated field, over-simplification carries many dangers. We would strongly suggest that any person who may want to truly start to understand the science of control and control systems should consider studying an engineering-oriented book on this topic.

Review Questions

1. What are the four basic components of a feedback control loop?

2. What are process loads and what impact might they have on a process?

3. What are the four basic process measurements?

4. What does PID stand for?

5. Briefly describe the three feedback controller actions.

6. What is the process set point?

7. What does it mean to "tune a loop?"

CHAPTER 4

Process Control Systems: A Theory of Evolution

Without getting too far out in the ether, the theory of evolution basically says man evolved from a very basic primate. What does that have to do with process control, you ask? Well, process control systems have had a similar type of evolution from the purely basic to an incredibly sophisticated level. In order to understand the architecture and functional layout of today's process control systems, it is useful to know how they evolved over time. While this will be a quick history lesson, it should be easy to see how the impact of decisions made in the design of early systems can be seen in control systems today.

The earliest and simplest automatic control systems were based on mechanical mechanisms, such as the mechanical link system found in a toilet. Figure 4-1 shows a simple automatic mechanical level controller. This mechanism includes a float that sits in the tank in which the liquid level is being controlled. The float is attached to a mechanical arm attached to a pivot arm, which is in turn connected to a fixed pivot and then connected to a valve mechanism. As the level in the tank increases, the float rises, causing the pivot arm to rise on the float side and descend on the valve side. The lowering of the pivot arm on the valve side causes the valve to close an amount proportional to the rise of the liquid in the tank. The closing of the valve reduces flow into the tank, which reduces the rate of rise of the liquid level. If the level goes down, the reverse takes place. This simple mechanical device does a fairly effective job of controlling tank level. It requires no power source and is inherently safe as a device in that it will not cause a spark that might ignite any flammable liquids. The primary disadvantage of such a control system is that it is only practical in very local environments. It cannot be managed remotely by an operator.

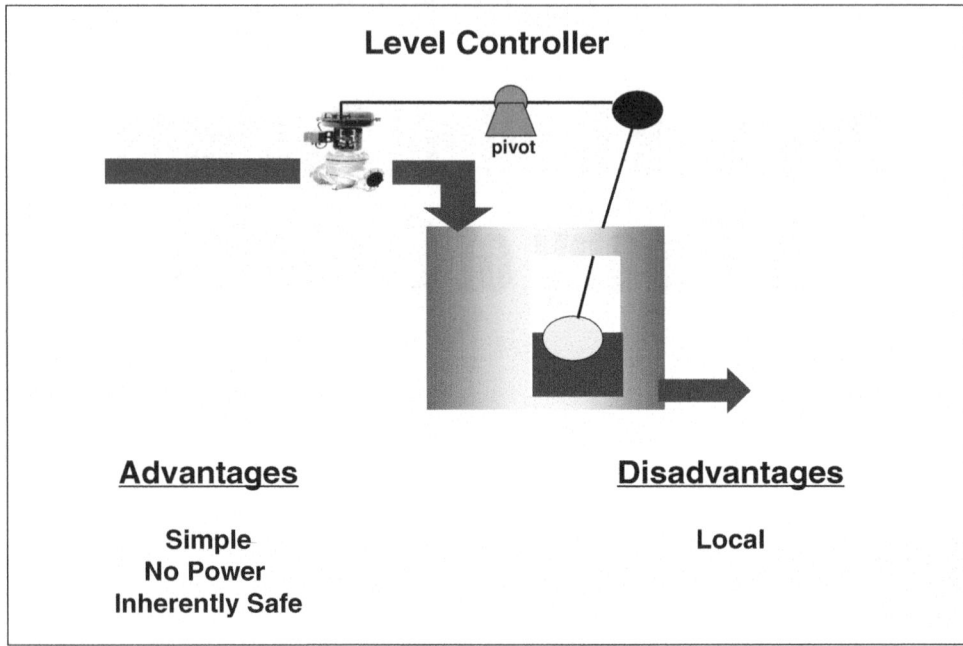

Figure 4-1 Mechanical Control System

After basic mechanical control systems came pneumatic control systems (Figure 4-2). Pneumatic control systems were the first powered systems. The power for these systems was air pressure. A common signal range of 3 to 15 pounds per square inch (psi) was agreed upon across industry for these systems. This means that the range of 3 psi to 15 psi represents the range of both the measurement device and the valve (which, for our purposes, may also mean other devices used to adjust the process) within the control loop. A 3 psi signal to a flow controller might mean that 0% of flow is going through the measurement device. A 9 psi signal might mean that 50% of maximum flow is going through, and a 15 psi signal might mean that 100% of the possible flow is going through the device. A 3 psi signal from a controller to a valve may mean to close the valve completely, a 9 psi signal may mean to open it to 50%, and a 15 psi signal may mean to open it all the way.

Using a standard signal range like this enabled multiple instrument, controller, and valve suppliers to include their devices in a pneumatic loop with equipment from other suppliers. In a sense this was the first standard fieldbus. The setting of the low range of the pneumatic signal at 3 psi rather than 0 psi enabled operators to distinguish between a low measurement condition and a system power failure. Engineers developed very clever pneumatic devices to be able to measure, control,

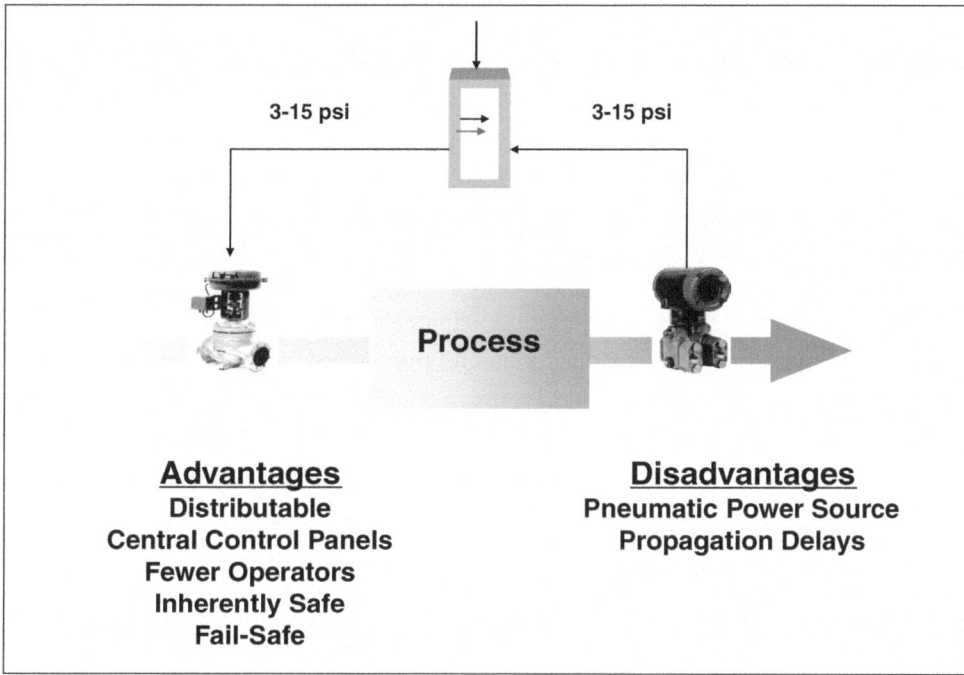

3-15 psi   3-15 psi

Process

<u>Advantages</u>
Distributable
Central Control Panels
Fewer Operators
Inherently Safe
Fail-Safe

<u>Disadvantages</u>
Pneumatic Power Source
Propagation Delays

Figure 4-2 Pneumatic Control — Air Power

and actuate with pneumatic power. The origin of PID control can be traced to these pneumatic devices.

Pneumatic control systems offered important advantages over the mechanical systems they were designed to replace. One of the greatest advantages was because the pneumatic signals could be transmitted over distance; the controllers could be located away from the measurement instrument and valves, enabling centralized control panels. Operators could supervise a number of loops from the control panel. This led to significant cost savings due to the reduced number of operators required to operate the plants. See, even back in the old days, they were trying to increase productivity.

As with mechanical systems, pneumatic systems are inherently safe because air power does not cause sparks that can lead to explosions. Also, because of the air-to-open and air-to-close design of pneumatic valves (as previously discussed), pneumatic control systems can go to a safe state with a loss of power.

Although there are considerable advantages with pneumatic control systems over mechanical control systems, there are also a few issues with these systems that must be (and certainly were) considered. First, the cost of the pneumatic power source for a plant can be quite high. Also, pneumatic signals do not propagate instantaneously; rather, there can be a time delay between the sending and receiving of a pneumatic signal that is proportional to the length of the transmission line, and any delay in a control loop introduces difficulties in controlling the process. Although pneumatic systems were a major advancement, these issues led control system designers to search for more effective approaches.

Pneumatic control systems use continuous signals ranging from 3 psi to 15 psi to represent the measurement across its range and the desired valve setting across the range of the valve. Data representation across a continuous range of signals, such as that represented by pneumatic signals across a 3- to15-psi range, is referred to as analog data. With analog data, the value of the signal directly represents the value of the variable being measured. Therefore, a pneumatic control system could be referred to as an analog control system.

To try to overcome some of the shortcomings of pneumatic control systems, engineers developed electronic analog control systems (Figure 4-3). These are also analog control systems, but they are based on an electrical power source instead of a pneumatic one. The accepted analog range for these control systems is 4 to 20 mA. Amperage is used (instead of voltage) for signal stability. The 4- to 20-mA signal range is an industry standard for electronic analog systems, as is the 3- to 15-psi signal range for pneumatic systems.

Electronic analog devices were able to perform the same basic functions as the pneumatic devices. This made it easy for an engineer with experience with pneumatic control systems to move to electronic analog control systems. Electrical signals can travel over much longer distances with essentially no propagation delays, making electronic analog control systems much more distributable than pneumatic systems. With these analog systems a process operator could manage a large number of control loops from a single location (a central control room), resulting in a reduction in the required number of operators. The operator headcount reduction led to cost savings sufficient to justify the expenditure on these systems. You can now see where the manufacturing industry was headed.

On the downside, electronic analog control systems introduced electricity into areas of process plants in which a spark might cause an explosion or fire. Devices called electrical barriers were developed and often had to go into the circuit to reduce the probability of a spark in areas of the plant where that would be a safety risk.

Also, since most valves were still pneumatic, plants that installed electronic analog systems typically had to support both pneumatic and electrical power sources in the plant, which presented added installation and maintenance costs. Finally, although pneumatic systems did not offer any advantage in this area, the calculation capability of electronic analog systems was very limited. These shortcomings led engineers to start looking for advanced ways to ensure that their plants were operating efficiently. The ability to do better, general-purpose calculations increased in importance. This became a critical limiting issue for analog systems.

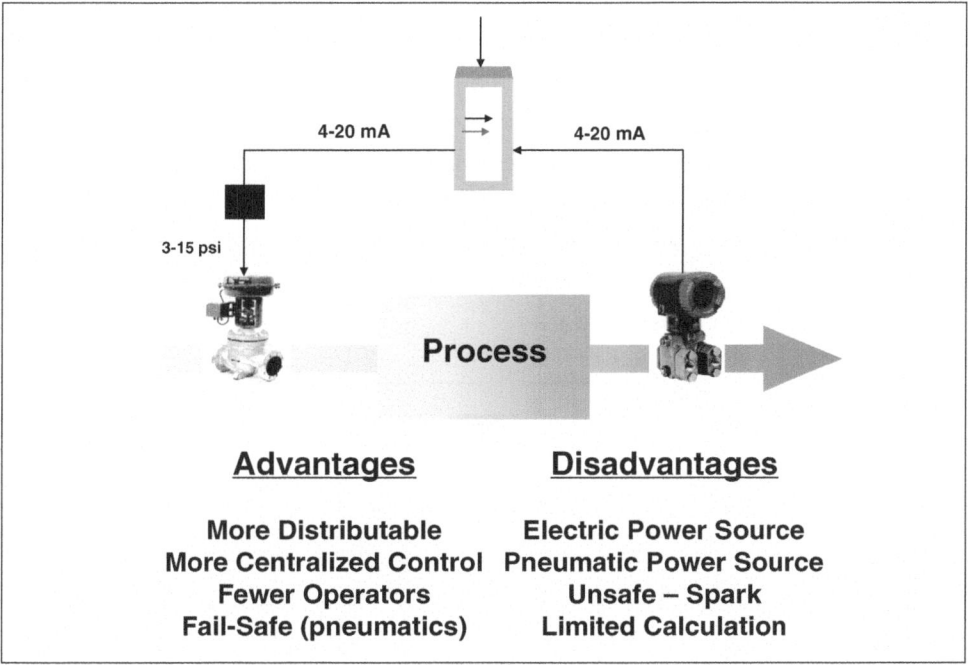

Figure 4-3 Electronic Analog Control

As their limited calculation capability started to become a significant issue for analog control systems, the digital computer age was beginning to show signs of life as prices started to come down to a level that made digital computation financially viable in manufacturing operations. Unlike analog systems, digital systems use electrical circuits to simulate the digits of a number. Instead of a signal representing a range, the signal represents the value of a digit within a larger number.

This resulted in much more flexible calculating engines than the analog systems could provide. Computers were still quite expensive and tended to not be as reliable as was required in manufacturing, so the first digital computer systems hit the floor to ensure set points of the analog controllers were optimally set. The resulting hybrid digital/analog systems were referred to as Set Point Control (SPC) systems (also referred to as Supervisory Control) or supervisory systems (Figure 4-4).

The computer offered the calculation engine the engineers had been looking for, and the analog controllers provided the actual feedback control mechanism. An interface between the computer and the set point input of the controllers was developed, and software in the computer was programmed to facilitate the calculation and downloading of the set points. Ensuring optimal set points enabled more effective and efficient plant operations.

These hybrid systems offered a level of distributed control, with the control done in the analog controllers, which meant the control would not fail if the computer went down. As we have seen, this was very important because the early digital computers were not very reliable and would fail frequently. Also, with computer displays and computer-based, set-point determination, operators could manage a much larger section of the plant than was previously possible, resulting in further headcount reductions in operations staff.

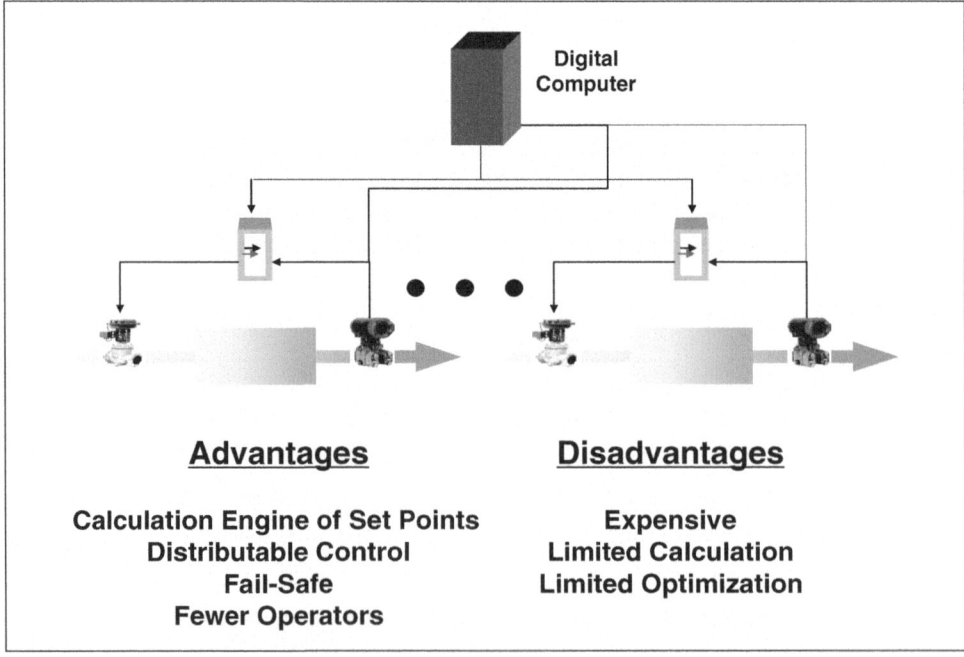

Figure 4-4 Digital Control — Set Point Control

On the downside, SPC systems were expensive because they required a digital computer and an analog control system, both of which were quite expensive by today's standards. The computer's calculation capability was limited to the functionality above the control level and did not penetrate the control layer. Also, in spite of the calculation capability of these systems, expertise and software for optimization were in their infancy and very little true set-point optimization was actually implemented.

With the disadvantages of SPC systems, a number of manufacturers implemented digital computers that could perform both control and advanced calculation functions. These systems were initially referred to as Direct Digital Control (DDC) systems (Figure 4-5). With DDC, system software in the computer replaced the control functionality previously provided by the analog controllers. Interfaces transferred measurement signals into the computers, and output signals from the computers to the valves. The signals from the measurement devices were often analog signals that had to be converted to digital signals for use by the computer. Suppliers developed analog-to-digital (A to D) converter devices to provide this function. On the output side, because valves were also designed for analog signals, D to A converters were developed and implemented. DDC systems offered the advantage of being less expensive than SPC systems because the analog control system was not necessary. They also offered unlimited calculation capability at the control and set-point optimization levels, allowing for creative control approaches.

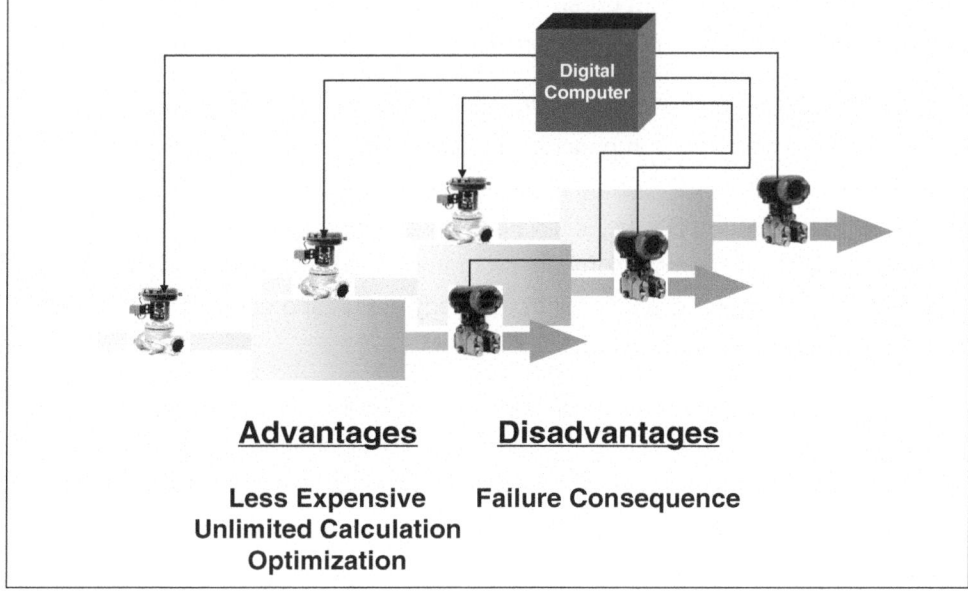

Figure 4-5 Direct Digital Control

The largest downside of DDC systems was the huge consequence of a system failure. Since digital computer technology was so expensive during this phase of the evolution of process control systems, the tendency was to put as much control as possible into a single computer, in spite of the risk. Some manufacturers tried a backup analog control system, but this solution was also very expensive. As the cost of digital computer technology continued to decline, and reliability increased in proportion, the direct digital control approach became the norm for the design of control systems.

One of the major challenges in moving from pneumatic and electronic analog systems to digital computers for process control was preserving the high-value knowledge of control engineers while undergoing this transformation. Control engineers were well trained in the development and implementation of control strategies with analog control systems.

Analog control systems were designed and developed through the selection of hardware components built to perform specific functions within a control strategy. For example, to develop a simple feedback control loop, an engineer might select an electronic analog instrument, a proportional and integral analog controller, an electronic analog-to-pneumatic signal converter, and a pneumatic valve. More complicated control strategies were developed by selecting the necessary additional hardware components and connecting them in the appropriate order. Control engineers developed significant expertise over years of working with analog systems, based on this hardware component approach.

With digital computers the control functionality goes through software rather than through the hardware components of analog systems. As digital computer technology hit the floor, few control engineers were versed in general programming languages. To deal with this issue, software development engineers at The Foxboro Company devised an ingenious software programming language to emulate the hardware component approach to control strategy design and development that the control engineers had been used to.

This was done through a software construct called the "Block Concept." Conceptual entities called software "blocks" were developed for each of the major functional entities required in an analog control system (Figure 4-6). Control engineers could literally build control schemes using the same knowledge they had developed with analog systems. They could sit at a computer terminal and work with configuration software that allowed them to build PID blocks, analog input blocks, analog output blocks, and a variety of other blocks, and connect them through software connections. It took very little time and effort using this software block approach to transition a good control engineer accustomed to analog control systems to digital control systems. The software block language invented for process control systems actually was an early form of object-oriented programming, which has become very important in computer programming over the past decade.

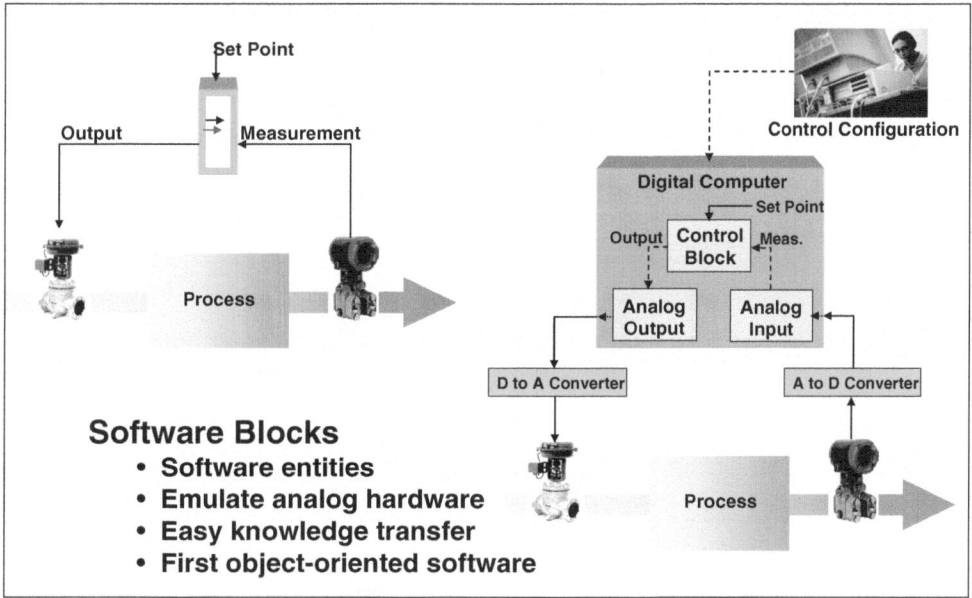

Figure 4-6 Process Control Software — The Block Concept

The reduction in cost-per-function of digital computer technology accelerated through the mid to late 1970s, to the point at which providing control functions in digital computers was actually considerably less expensive than providing the same functionality in analog systems. This, combined with the acceptance by industry of block programming languages and direct digital control systems, naturally led to the evolution of a new class of systems (Figure 4-7), distributed control systems (DCSs). Honeywell introduced the first DCS, TDC2000, in the late 1970s, setting a new direction for control system design that would take the industrial automation through the latter decades of the twentieth century.

DCSs offered a number of advantages over the previous digital control systems. First, DCSs are much more distributable than previous digital control systems were. They are distributable in two important ways: geographic and functional. In geographic distribution, the computer modules of the system are networked together and distributed to different locations in the plant. In functional distribution, the computer modules are designed to perform a specific function within the overall function set. For example, one computer module of a DCS might be dedicated to the function of process control while another might be dedicated to the function of operator interfacing. The combination of geographic and functional distribution provided by DCSs resulted in extremely flexible system designs that could match a wide variety of requirements across a wide variety of industries.

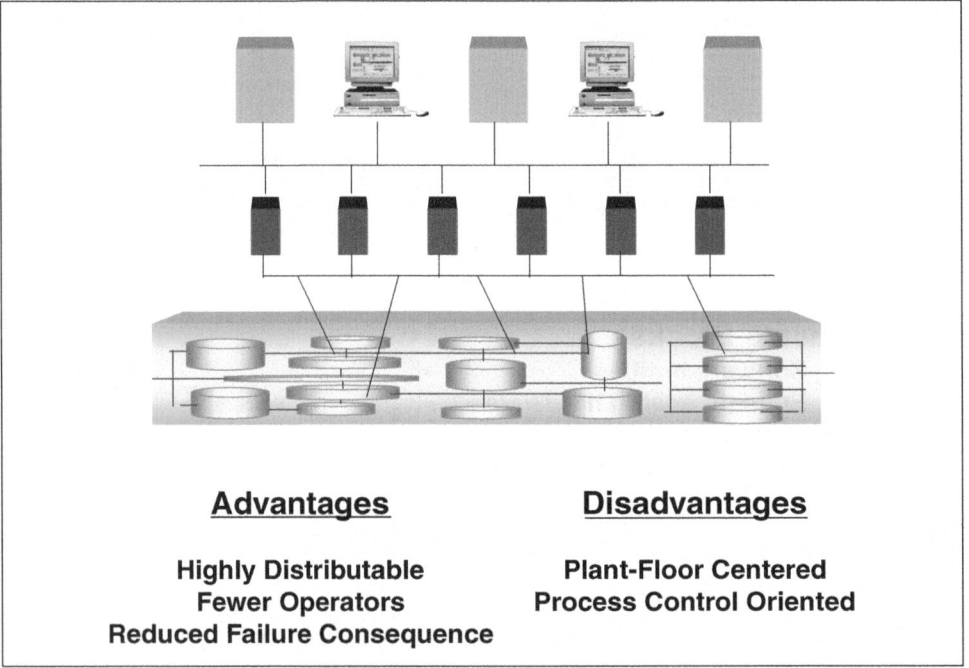

Advantages

Highly Distributable
Fewer Operators
Reduced Failure Consequence

Disadvantages

Plant-Floor Centered
Process Control Oriented

Figure 4-7 Distributed Control System

Second, DCSs typically also allowed much more centralized supervision over much larger sections of process plants, often resulting in single control rooms for very large plants. This resulted in further cost savings due to a reduction in operators. Third, by distributing the functionality across a number of computer modules, the consequence of a failure in any one module was much more contained. In addition, DCS suppliers designed redundancy and fault tolerance into critical modules in the DCSs so the failure of a component would not result in any loss of function.

Perhaps one of the major disadvantages with DCS designs is that, although they are based on digital computers that are very similar to those used in business computing systems within the same plants, DCSs are so process-control and plant-floor oriented that their design has actually tended to discourage interoperation with the business systems. This separation between DCSs and business systems has been an issue ever since the introduction of DCSs, and it is one of the top issues facing the automation industry today.

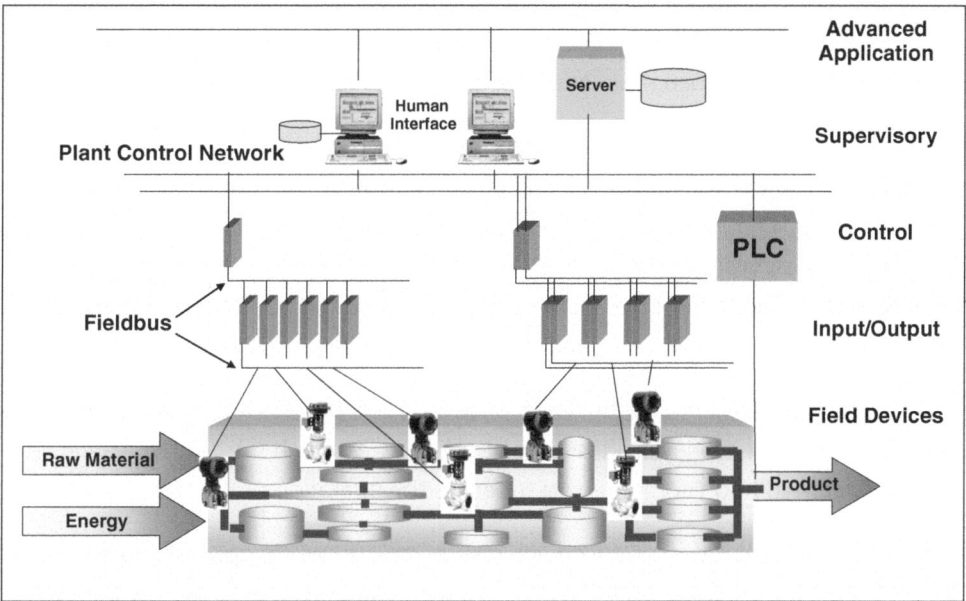

Figure 4-8 Distributed Control System — Basic Architecture

Although DCSs have been designed by a number of different automation suppliers, such as Honeywell, Foxboro, ABB, Siemens, Yokogawa, and Emerson, to name a few, over time their basic architectures have converged to the point at which most DCSs today share a fairly common architectural design that can be described as five layers (also called levels) of functionality. Figure 4-8 presents the five-layer architecture that represents the common DCS design found across most suppliers' systems today. Starting at the lowest layer, instruments and valves have traditionally been considered to be outside of the DCS scope, but with the increased availability of digital intelligence within these (and other) field devices, they are starting to be considered part of the control system.

With intelligence in the field devices, the importance of a digital communication infrastructure to connect them to the DCS has become apparent. The resulting digital networks at this level have been classified as fieldbuses. A number of proprietary vendor-developed fieldbuses were originally introduced to the marketplace, but they only allowed the field devices to interoperate with the systems if they were provided by the same vendor. An industry standard fieldbus was developed by the Fieldbus Foundation, a digital fieldbus standardization organization, in an attempt to enable intelligent field devices from any vendor to work with DCSs from any vendor. The development of a standard digital fieldbus required considerably more effort than that required for the analog fieldbuses because the addition of intelligence to field devices meant that there was much more information of interest to the DCS.

The next layer in the architecture is the input/output layer. This layer provides the interface to the intelligent and non-intelligent field devices. Even though there are advantages to having intelligence in the field devices, there are millions of devices, installed in process plants prior to the availability of intelligent field devices, which are still there and will be there for years to come. Intelligent and non-intelligent devices need to be brought into the DCS, and the input/output layer addresses this.

The next layer up is the control layer, at which basic process and logic control occur. Above the control layer is the supervisory layer, at which the operators interact with the system through computers to supervise the operation of the plant.

The highest level is the advanced application layer, at which applications such as process historians, optimizers, and production planning and scheduling run. DCSs may not be physically partitioned into these layers, but the functionality within these systems is often partitioned this way. Thinking of DCSs in this way helps when trying to align required functionalities within an overall architecture.

The transition to today's DCSs is the culmination of an evolutionary process over the last century (Figure 4-9). Gaining an understanding of this progression over time can be helpful when trying to understand the design of today's systems. Some of the design components, such as the block structure for system configuration, arose from the requirements of earlier generation systems.

Even though there have been advances in digital control system design, some of which will be presented in later chapters, DCSs still represent the current state-of-the-art in control systems design.

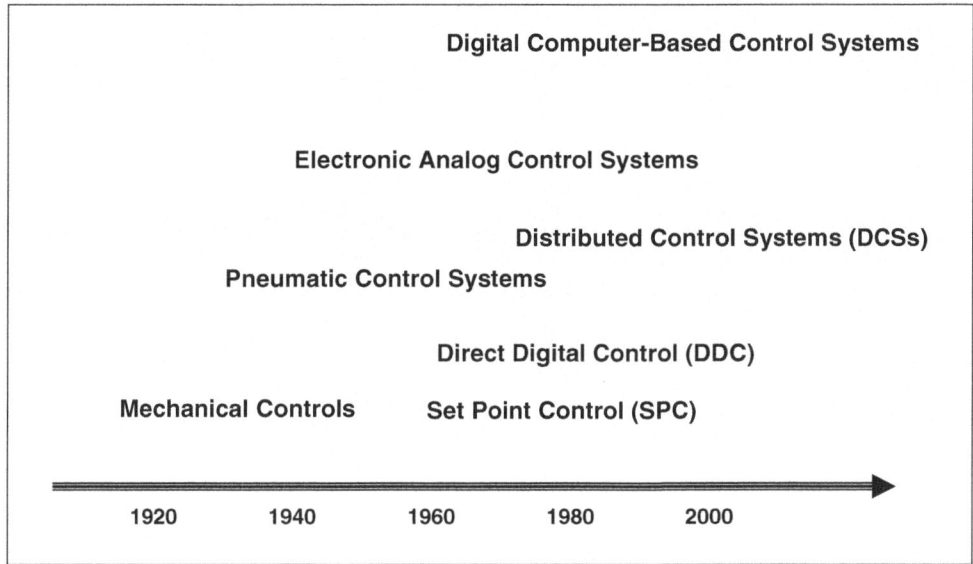

Figure 4-9 Control System Evolution

Review Questions

1. Name a simple everyday example of a mechanical control system.

2. What is the power source in a pneumatic control system?

3. What is the standard signal range for a pneumatic system?

4. What is an analog signal?

5. What is the standard range for signals in an electronic analog control system?

6. Briefly describe a set-point control system.

7. What does "direct digital control" mean?

8. Describe the function and purpose of the software block structure.

9. What is the purpose of a fieldbus?

CHAPTER 5

Control of Batch Processes: Let Them Eat Cake

Your birthday is soon approaching and you are looking forward to the big day. Why? Not because you are eager to celebrate yet another advancing year, but rather you always get to gobble down your favorite cake. It is the greatest cake of all time and you only have it once a year. Every mouthful is pure enchantment of rich flavor and moist texture. It tastes perfect every year.

Your birthday cake is a perfect example of a batch process at its best.

Just think, a recipe defines the steps required to prepare and bake the cake correctly (the procedure) and the formula—the quantities of ingredients, such as eggs, milk, and flour, and the other variables, such as oven temperature, baking time, and cooling time. Both the correct procedure and the correct formula are necessary to produce the desired slice of heaven on your birthday.

A batch or discontinuous process is what takes place when a manufacturer makes products in batches or lots as compared to continuously producing them. Although any product made through continuous processing could be produced via batch processing, the products made in batches are typically those for which either a degree of manufacturing flexibility is beneficial, or for which at least one stage of the manufacturing process requires an extended amount of time to complete. An example of an operation that may benefit from a degree of flexibility is bulk pharmaceuticals, where a manufacturer often makes dozens of products in moderate to small quantities. It is most economical to produce such products using the same process equipment. Since the market can only support limited volumes of each product, producing high volumes of any single product would not make economic sense.

One operation in which at least one step in the manufacturing process requires an extended amount of time to complete is making latex paint. The primary chemical process involved in the making of latex paint is an exothermic (heat-producing) chemical reaction. In paint manufacture, chemicals mix in a vessel, in this case a reactor, and a chemical catalyst is charged to the mixture. Because of the properties of the catalyst, it chemically reacts with the mixture in the reactor, and heat results. Once the catalyst is charged to the reactor, the chemicals react over an extended period of time until the desired product remains in the reactor vessel. During the reaction, the ingredients are mixed to ensure uniformity, and when the reaction is complete, the product cools to the desired temperature. When the product has cooled, the manufacturer removes the product from the reactor for further processing. The extended nature of the reaction would make it difficult to manufacture these products continuously.

In batch manufacturing, the products are produced through a sequence of manufacturing stages, typically called the phases of operation. Each phase of operation typically consists of a number of sequential or parallel manufacturing steps. At the basic control level, batch manufacturing operations involve logical control activities, such as turning on and off pumps or agitators, and process control activities, such as controlling temperature or level. Therefore, basic control in batch manufacturing operations typically involves a combination of process controllers and programmable logic controllers. Because of this combination of control approaches at the basic control layer, batch control is sometimes also called hybrid control.

Although some batch operations consist of the repeated manufacture of the same product, most involve producing multiple products or multiple grades of product through the same process equipment with the same basic logic and process controllers (Figure 5-1). A recipe that defines the specific manufacturing requirements for a product is developed for each product or product grade to be produced through a specific set of process equipment. As we have seen, recipes have two basic components: procedures and formulas. The procedure defines and controls the sequence of phases and steps required to make a specific product through the process equipment. The formula defines the materials, quantities of materials, times, levels, temperatures and other variables that will be used per the procedure to match the product requirements to the manufacturing equipment in which the product is to be made. Again, the perfect example is the cake concept mentioned earlier.

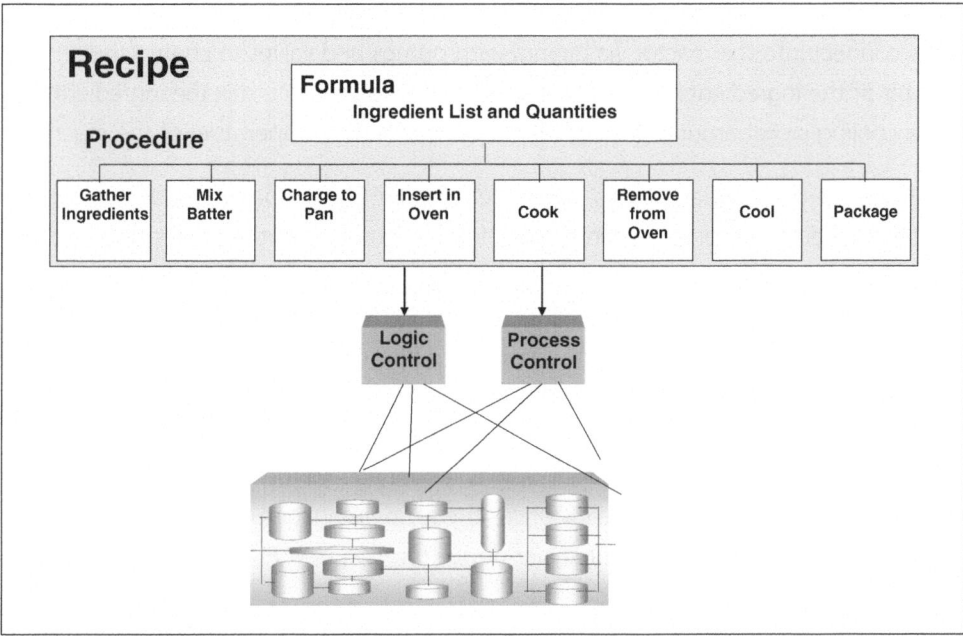

Figure 5-1 Batch Process Control

Basic batch process control tends to be more complex than either basic process or logic control. The primary reason for this is developing even a minimal level batch control system first requires the implementation of process control and logic control, then the implementation of a coordination level above the basic control level in the form of recipe management. Also, as we have seen, formulas and procedures often have to be developed for each product or product grade being produced. Control engineers can typically approach continuous process control challenges a loop at a time and discrete process logic control challenges a work cell at a time, but batch manufacturing operations must be coordinated and controlled as a whole. This makes the control of batch processes more of a system analysis and development problem than just a control problem. The engineering talents required for successful batch processing can be quite different from those required for continuous process and discrete logic control.

As was previously discussed, processes involving exothermic chemical reactions are often developed as batch manufacturing processes. Making latex paints is one example (but note that batch processes may also involve endothermic reactions, such as the baking of a cake). In Figure 5-2, there are three ingredient tanks containing the basic chemicals that go into making the paint. There is a tank containing the chemical catalyst that will cause the ingredients to react when combined. These four storage tanks connect into the reactor via piping, with pumps and valves to control the transfer of the ingredients. The reactor vessel has an agitator to mix the ingredients and a cooling jacket around the outside that controls the temperature of the reaction.

There are three storage tanks, one for each color of latex paint produced. After a batch of product is completed in the reactor, it is pumped to the appropriate product storage tank. The products in the storage tanks are pumped through piping to a transportation area where they are loaded into trucks that transport them to a packaging facility in which the paint is packaged into cans. The control system in a paint plant might consist of a distributed control system (DCS), a group of process controllers, and some programmable logic controllers (PLCs). The recipe management software would typically run in the DCS, which connects to the process controllers and the logic controllers. In some instances the recipe management software may operate in a personal computer or server device running over PLCs.

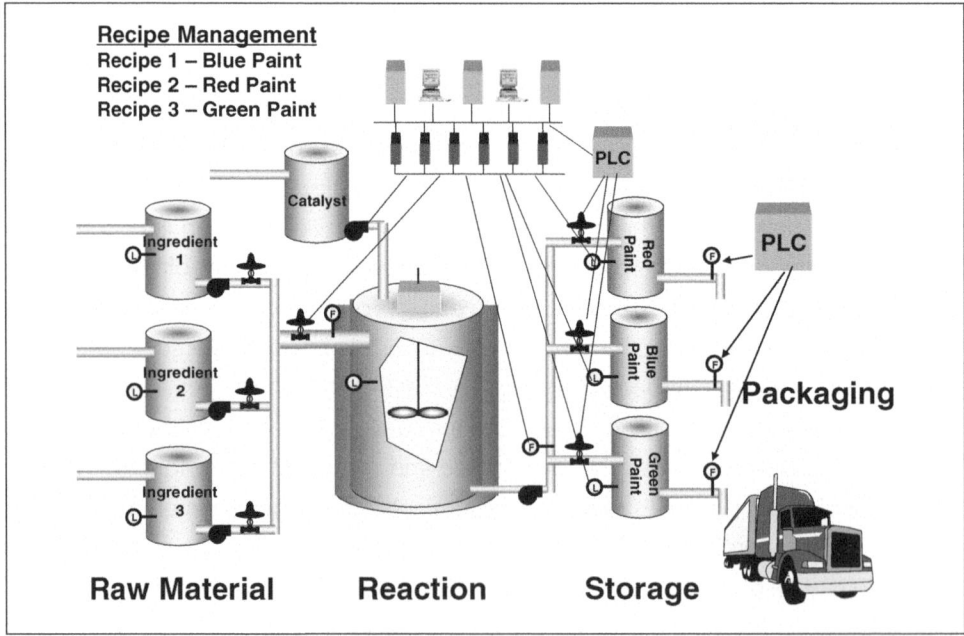

Figure 5-2 Batch Process Example — Exothermic Reaction

This example helps to demonstrate the holistic nature of the control of batch processes. If one control system were controlling the ingredient storage tanks while a different system controlled the reactor, and a third system controlled the discharge into the product storage tanks, it would be extremely difficult to coordinate the production of a batch of paint across the plant. Notice also that in the example the PLC controlling the loading of finished product into the trucks is not connected to the DCS. This is because the loading of the trucks can be effectively accomplished independently from the making of the products as long as the batch control system has access to the level measurement of the product storage tanks to make sure there is room in the tanks for the finished products.

Understanding normal sequence of operations for batch control is often very straightforward. At least this is the case if everything goes according to plan. One issue that can crop up with batch control software is how to respond if something goes wrong in the production of the batch. Suppose during the reaction phase of the latex paint, the reactor cooling system fails in some way. In this situation there will be an exothermic chemical reaction that continues to produce heat, and the mixture will get hotter and hotter until it could explode.

Clearly it is very important for the control system to effectively address exceptions of this kind. Batch control software is partitioned into normal control logic, which defines the normal operations involved with the manufacture of a batch, and exception logic, which defines what action the system will take on the identification of an exception in the processing (Figure 5-3). The exception logic runs in the software of the batch control system and monitors the process in order to identify the existence of an abnormal event.

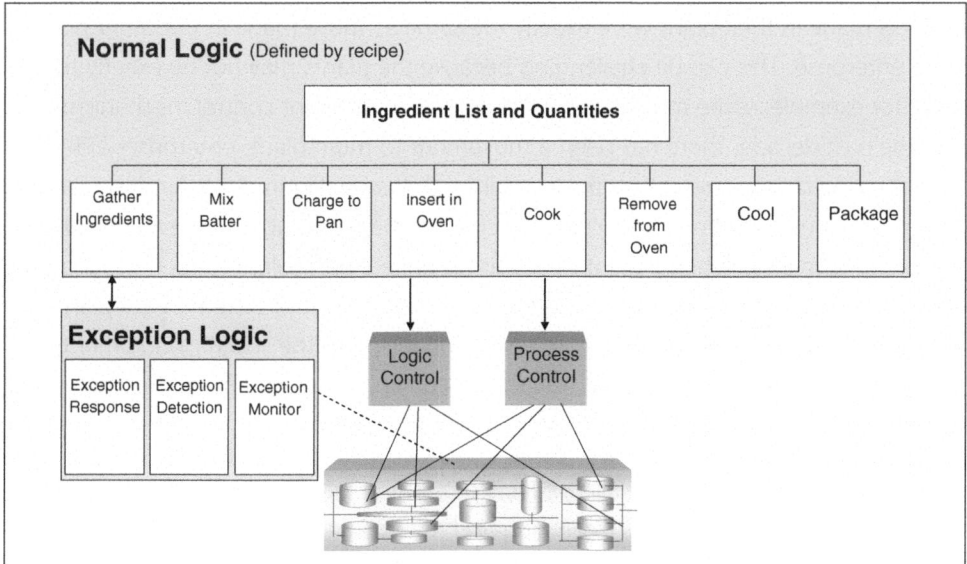

Figure 5-3 Batch Process Control with Exception Logic

The logic must be intelligent enough to know when an operation is normal and when it is not. For example, the only time the reactor cooling system has to operate is when the reaction is taking place. When the ingredients are being charged to the reactor, or the product is being transferred from the reactor to the storage tanks, the cooling system should be turned off. It is only during the reaction phase that the cooling system must be operating and when its failure to be operating is considered an exception. The normal logic must coordinate closely with the exception logic to identify the true exceptions.

Once an exception has been identified, the exception logic has to determine the appropriate response. There are typically two classes of exception responses; they are service and hold. A service response is the less critical of the two and typically includes a pause in the normal logic and notifying the operator of the situation. The operator will then review the exception and decide how to proceed. If a more critical exception is identified, and a hold response is required, the exception logic will attempt to automatically drive the process to a safe condition. The safe condition is a function of the exception and the phase of operation the process is in. The hold response to the exception logic provides a built-in safety system for batch manufacturing processes.

Companies that have batch manufacturing operations often have multiple plants that make the same products. Producing consistent products in different plants across an enterprise can be important to the marketing of the products and very challenging. Historically it was the responsibility of the plant operations team to make sure the products made in their plant were exactly the same as those made in the other plants in the enterprise. This can be challenging because the plants may not be exactly the same; for example, some may have smaller vessels or different control mechanisms. Over the past decade, there has been a movement to multi-plant, enterprise-wide general recipe management systems to address this issue (Figure 5-4). A general recipe file is maintained to define the accepted recipe for the manufacture of every product the company makes. When a plant needs to produce a particular product, the master recipe transfers from the master recipe file to the batch control system in the plant. The plant's batch control system has to adapt the recipe to the specific equipment configuration in the plant and then execute the batch. In this manner the required level of enterprise-wide product consistency can be maintained.

The batch control software also collects processing information for every batch produced in a batch history file. Collecting and maintaining batch history was originally done just to monitor the process and to be able to understand which differences in processing resulted in the best products. But many pharmaceuticals and other human consumables are produced in batches, and collecting specific data on how a particular batch was produced and storing it for later analysis has become very important when a problem occurs in which the end product suffers. The batch history data allows analysts to review how the batch was made to try to determine how the product was corrupted. The unique batch identifier associated with each batch of product produced can be utilized to determine where else material from the corrupted batch may have been used or may have gone.

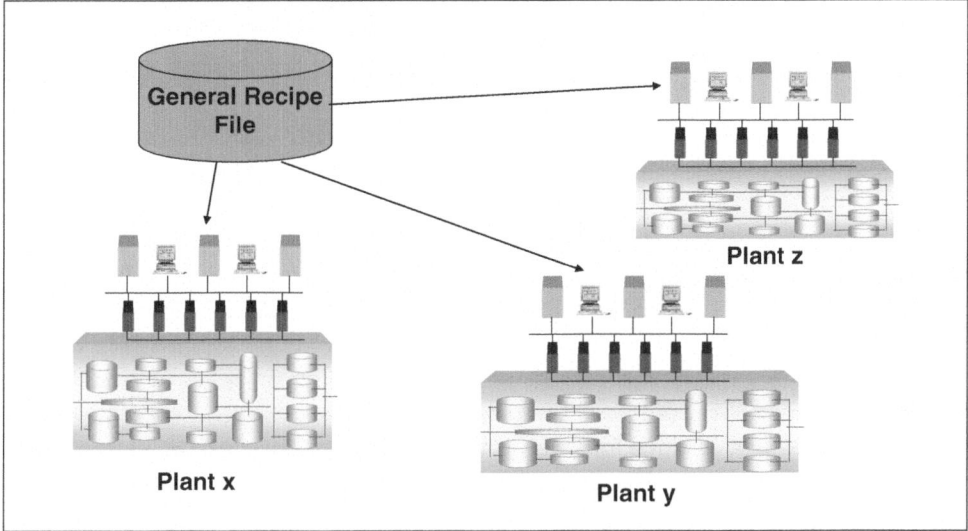

Figure 5-4 Multi-Plant Recipe Management

As has been discussed, batch processing operations must produce multiple products or multiple grades of product. Maximizing the business value from these operations is a function of how effectively they operate, but it is also a function of how effectively decisions occur as to which products to produce at any time. In most multiproduct batch operations, there is a layer of software that runs over the plant's recipe management software to determine the optimal production schedule—which products and how much of each to make. This software is commonly referred to as planning and scheduling software (Figure 5-5). There are numerous schemes and approaches for determining the optimal schedule of products to produce. Today, some manufacturers determine the schedules for all their plants from a central planning and scheduling function. This allows them to optimize shipping costs, maximize product shelf life, and meet market demands more flexibly while also managing effective batch operations.

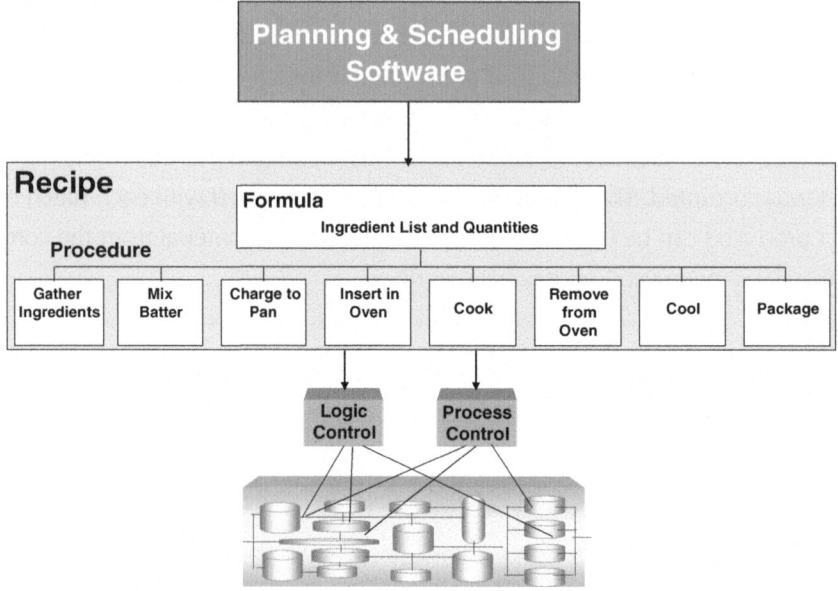

Figure 5-5 Batch Planning and Scheduling

Controlling batch manufacturing processes certainly presents different challenges to manufacturing companies than controlling continuous or discrete manufacturing operations. Batch process control combines the technological approaches of continuous and discrete process control, but adds a layer of complexity due to the fact that batch processes must be managed as complete entities. Quite a few of the more advanced concepts associated with batch control are starting to appear in continuous and discrete manufacturing operations as the level of sophistication required to optimize these operations increases. Let them eat cake.

Review Questions

1. What are the most common basic control elements for batch control systems?

2. Name and briefly describe the two basic components of a batch recipe.

3. Batch processing is divided into a sequence of processing stages that are typically referred to as what?

4. The logic built into batch control software that enables effective responses to be developed if something goes wrong in the processing of a batch is called what?

5. Enterprise-wide recipe management is coordinated through a recipe storage and distribution mechanism called what?

CHAPTER 6

Advanced Process Control: Beauty Beats the Beast

A s beauty is in the eye of the beholder, if you want to take the analogy to an extreme, so too is advanced process control.

The phrase "advanced process control" (APC) means different things to different engineers, and some of the schemes discussed here may not always be considered advanced process control strategies. But for the purposes of this discussion, we will consider any control scheme that involves more than single-loop feedback control as advanced process control. We should note that some of the following advanced control schemes that involve more traditional control approaches are sometimes referred to as advanced regulatory control, while some of the more sophisticated may be classified as advanced process control, but for the purposes of this discussion, we will not distinguish between the two.

As the world has become more complicated, it is no surprise that process control has followed suit. That is why over the years, control engineers have expanded on single loop feedback control to develop more advanced control schemes designed to address specific control issues or expand the scope of control beyond any single process measurement. While at first there was trepidation in trusting the technology, today's demanding manufacturing environment is almost forcing the issue.

Cascade Control

Cascade control, which combines two control loops into a single control strategy, is a very common multi-loop control approach. Figure 6-1 provides a diagram of a cascade control strategy for controlling a heat exchanger that involves connecting the output of one controller to the set point of a second controller. The two loops in such a control strategy are the primary and secondary loops. The primary loop controls the primary measured variable and provides the set point to the secondary loop. In the heat exchanger example, the intention of the control strategy is to control the temperature of the water coming out of the exchanger. Therefore, the outlet water temperature is the primary measured variable. The set point of the primary controller is set to the desired water temperature.

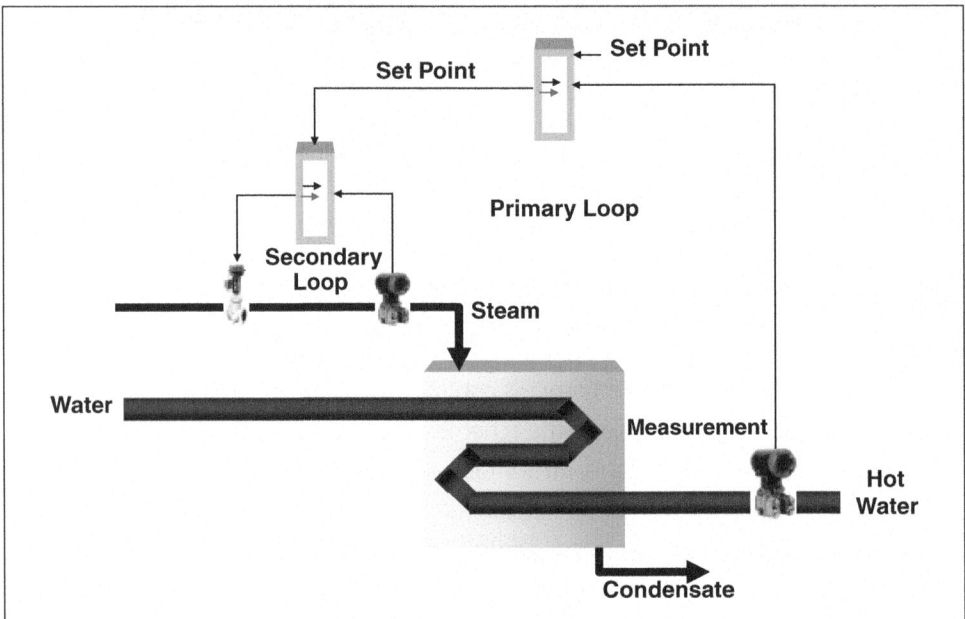

Figure 6-1 Cascade Control Example — Heat Exchanger

The outlet water temperature is controlled, in this example, by adjusting the steam flow into the jacket of the heat exchanger. The problem is steam is used by a number of different process units throughout the plant, and as these units are turned off and on, the pressure in the steam supply fluctuates considerably. These changes in steam pressure are a process load (disturbance) on the process. As you will recall, a process load is a variable that impacts the process, causing a measurement to change. In this example, the steam pressure becomes a process load.

It is critical to the success of the overall water temperature control strategy to regulate the steam flow at the inlet to the exchanger, which is accomplished by measuring the steam flow rate and opening or closing a valve. Since steam flow is not the objective of the overall control strategy, this loop is considered to be the secondary loop of the strategy and the output of the temperature controller is connected to the set point of the steam flow controller.

Cascade control is a common control strategy that can improve control stability significantly over single loop feedback control. But cascade control is only effective if the secondary control loop responds much faster to a set-point change than the primary loop does.

Feedforward Control

As we have seen, single loop and cascade control are examples of feedback control strategies. In feedback control, the controller waits until the measured variable is out of alignment with the set point before responding. One of the major shortcomings of any feedback control system is an error must occur before any corrective action is taken. An alternate and more advanced approach is feedforward control.

A feedforward control strategy is one in which the process variable to be controlled is measured along with as many of the process loads as feasible (Figure 6-2). For the heat exchanger example, the steam pressure, steam temperature, ambient room temperature, temperature of the inlet water, and minerals in the inlet water are all process loads. A change in any of these variables could lead to an error between the measurement and the set point. In most feedback control systems the process loads are not measured and changes in load resulting in a change in measurement are corrected after the fact. With feedforward control, the loads are measured and a mathematical model is developed that predicts what impact a load change will have on the measured variable. When a load change is detected, the feedforward model calculates a corrective action and transmits a signal to the valve controlling the controlled variable before any change in the measured variable is actually detected.

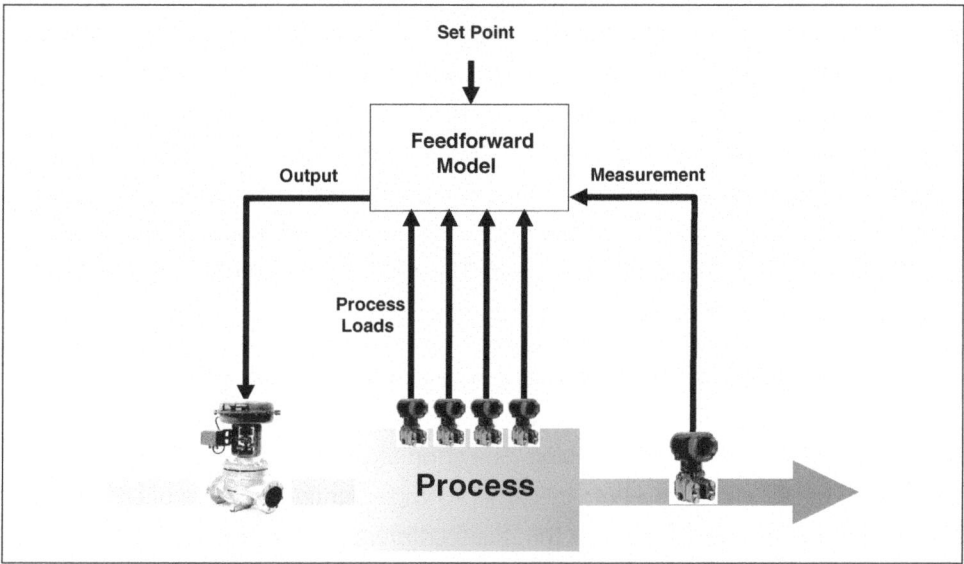

Figure 6-2 Feedforward Control

Since manufacturing processes do not change instantaneously with a load change, feedforward models typically require dynamic components to match the output signal to the valve with the dynamics of the process itself. For example, if a particular load change will take 10 seconds to impact the process in a way that affects the measured variable, and the manipulated flow will have an immediate effect on the process, the output to the valve may need to be delayed by an equivalent amount or else the valve will be adjusted too early, which will also lead to an error. If, on the other hand, there is a 6-second delay for the manipulated flow to impact the process, the output to the valve should only be delayed by 4 seconds. Careful consideration of process dynamics is essential for operation of feedforward control strategies.

Simple manufacturing processes have two basic dynamic responses: process gain, dead time, and lag. Process gain is the ratio of the change in the output of the process to the change in the input that caused it. Dead time is a process delay between a cause and an effect due to a process condition such as a mechanism's response time. For example, suppose a chemical is to be precisely charged to a vessel, and the mechanism installed to accomplish this is a measuring cup that has to be precisely filled prior to charging. Once an event occurs that would require the chemical to be charged, it might take 30 seconds for the mechanism to fill the cup and the chemical in the cup to be added to the vessel. The dead time between the detection of the event and the charging of the chemical would be 30 seconds.

A process lag is also a delayed response of the output of a process caused by a change in input to the process, but in this case the delay is a ramping response typically due to the capacity of the process. For example, consider the case of a large vessel containing a liquid that is to be heated to a specified temperature by increasing the gas flow to a heating element located below the vessel. This is similar to heating a pot of water on a gas stove in the kitchen. To cause the control action, the gas flow to the heating element is increased to the point required to cause the liquid to heat up to the desired temperature, but the liquid does not immediately jump to the end temperature. Rather, the liquid heats up to the desired point gradually over a period of time, almost in a ramping manner. This type of delay in reaching the desired temperature represents a lag in the process.

Building a feedforward control model can be challenging. There are always errors in the parameters and missing measurements. Getting the process dynamics right in the model can be even more challenging. The effort required to effectively build a feedforward model has been a primary barrier to implementing this type of control strategy. But a second issue has been that the feedforward model represents the conditions in the process at the time the model was developed. As a human physically changes over the years and is unable to run a 6-minute mile after the age of 40, processes also change over time. The change could be from wear or the buildup of minerals in the pipes or any number of other causes. When these changes occur, the feedforward model no longer represents the process correctly and loses its effectiveness.

Engineers can try to readjust the model to match the conditions, but the amount of work and the frequency of changes required make this impractical. The approach engineers normally take is to insert a variable into the feedforward model that can compensate for the changes (Figure 6-3). A simple and sustainable solution to this problem can be provided by incorporating a feedback controller in the feedforward strategy to adjust this variable in the feedforward model. This feedback adjustment is referred to as feedback trim, and adding feedback trim to feedforward control strategies helps to make feedforward control approaches much more practical in operating plants.

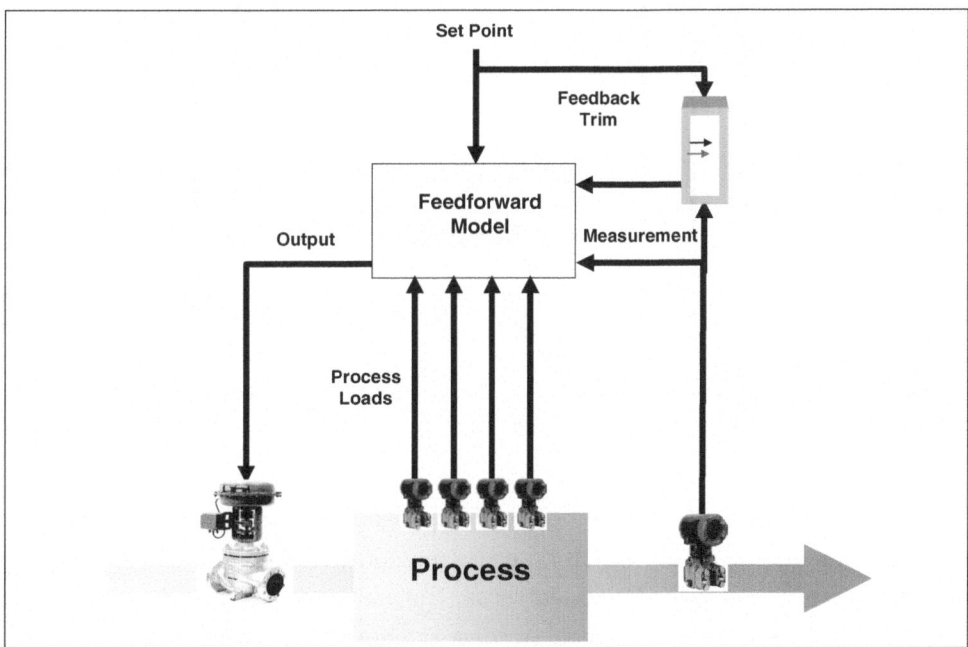

Figure 6-3 Feedforward Control with Feedback Trim

Multivariable Predictive Control

Although a number of other control strategies have been deployed in manufacturing process operations over the years, the final advanced process control strategy we want to look at is multivariable model predictive control (MPC). The essence of an MPC strategy is multiple process measurements; the outputs are controlled simultaneously through the implementation and execution of a process model (Figure 6-4). The process model evaluates the current value of the measurements and uses those values to determine the correct position of the outputs. MPC is similar in concept to feedforward control, but it extends the feedforward models and strategies to provide multiple coordinated process outputs.

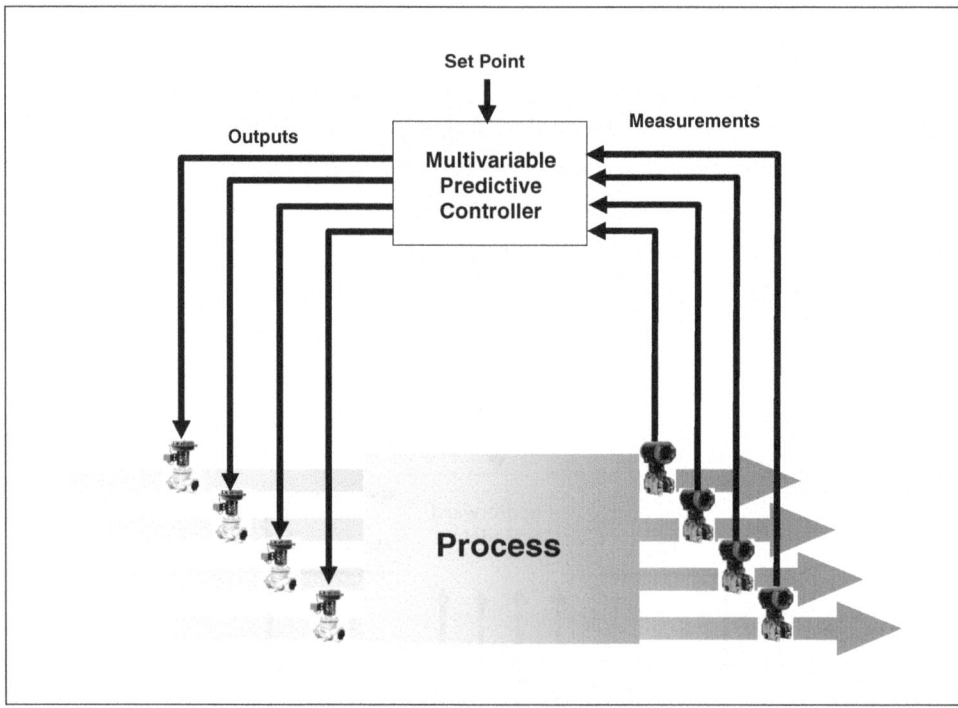

Figure 6-4 Multivariable Predictive Control

The primary approach used in the construction and execution of MPC models is the use of matrix analysis, which is based on measured responses and is not a mathematical model in the traditional sense. The matrix models use advance mathematical processes from the field of linear algebra to determine what the desired output settings should be for any given set of process measurements. As with feedforward control, process changes over time may result in models that no longer exactly reflect the current state of the process. To deal with this, some MPC controllers have built-in adaptive features that adapt the models in a somewhat similar manner to that of the feedback trim mechanism of feedforward controllers, and all employ some form of feedback correction.

Process manufacturers have had significant success over the past few decades in applying and maintaining advanced process control strategies. Although these advanced control approaches offer significantly improved control over single loop feedback controllers, the effort required to develop and maintain these strategies has limited their application to only those areas that are either very difficult to control with single loop feedback control or in which more stable and tighter control offers a significant economic benefit. Therefore, advanced process control strategies of these types have until recently been limited to a small percentage of the process control needs in manufacturing operations. To make matters a bit worse, in many of the operations using advanced process control strategies, process operators often feel uncomfortable with the controllers and end up switching them off and operating the process manually, which almost guarantees a loss in productivity. Certainly this has limited the potential positive impact that could result from the application of these strategies.

On the positive front, in today's manufacturing environment, where increases in productivity are a must to achieve greater profits, engineers and operators have developed increased confidence in advanced process control, and the percentage of advanced process control systems being used is on the rise.

Review Questions

1. What is the basic difference between single loop feedback control strategies and advanced control strategies?

2. What is the name of a control strategy involving two feedback control loops in which one controller outputs the set point to the second controller?

3. What is the name of a control strategy in which the measured variable and the measures of the process loads feed a model that determines the most appropriate setting for a valve to compensate for an upset in the process before the upset has a chance to create an error?

4. What is feedback trim and where is it commonly used?

5. A control strategy based on multiple measurements of the process feeding a model that determines multiple valve settings (outputs) simultaneously is called what?

CHAPTER 7

Optimization: Math Gone Wild

What may seem sensible in an everyday work environment sometimes just does not add up. That is why optimization software comes in pretty handy on the plant floor. It's all in the math.

Optimization software uses advanced mathematics to help determine the best possible solution to a problem. Optimization software can determine such things as optimal set-point settings for manufacturing processes, optimal production schedules, or optimal maintenance schedules. Currently, optimization problems with a single objective and multiple constraints can be resolved. The objective, or the optimal solution, is either the maximum or minimum of some desired outcome, such as maximum profit or minimum cost. The constraints can be the maximum physical capability of a section of a process, such as a pump, or a safety condition you cannot exceed, or a condition that might damage the equipment. In order to apply optimization software, you must be able to describe the objective and constraints using mathematical equations and inequalities. These equations and inequalities are the objective functions and the constraint functions, respectively. For example, if the temperature of a vessel must not exceed 520 degrees, the constraint function for this constraint is $x \leq 520$.

Simple Optimization

While optimization can be very complex, this simple example can provide some clarity. Suppose there is a paint plant that can make 100 drums of either red or white paint every day. The profit on the red paint is $60 per drum and the profit on the white paint is $40 per drum. The objective of the plant is to make the correct combination of red and white paint to maximize profitability. Also, suppose there are no constraints on how much of each type of paint can be made. The only constraints for this example are the plant cannot produce negative amounts of paint and the maximum number of drums of paint the plant can produce per day is 100. In this simple example, the objective function is:

Maximize $Z = 60x_1 + 40x_2$

Where:

Z is profit

x_1 is the number of drums of red paint made and

x_2 is the number of drums of white paint made

The three constraint functions for this problem are simply:

$x_1 \geq 0$

$x_2 \geq 0$

$x_1 + x_2 \leq 100$

In this simple example, it is easy to decide the plant should produce as much red paint as possible since red paint is more profitable than white paint, and there is no constraint on what percentage of the paint is red and what percentage is white. The optimal solution is to make 100 drums per day of red paint and 0 drums per day of white paint. Figure 7-1 provides a graphical representation of this problem. The x axis represents the constraint $x_1 \geq 0$ and the y axis represents the constraint $x_2 \geq 0$. The line at the upper edge of the shaded region on the graph represents the constraint $x_1 + x_2 \leq 100$. The shaded area on this graph represents the possible solutions to the problem.

Although this is a fairly trivial example of an optimization problem, it is useful in pointing out some important characteristics of these problems. First, there are an infinite number of solutions that could satisfy the requirements of the constraints, as represented by the shaded region in the graph. Most of this infinite number of solutions would not meet the objective of the optimization, which in this case is to maximize profit. The optimal solution to all problems of this kind will always be on the edge and at a corner point of the region of feasible solutions. In this case, the lower right vertex of the triangle, which represents producing as much red paint as possible and no white paint, is the optimal point.

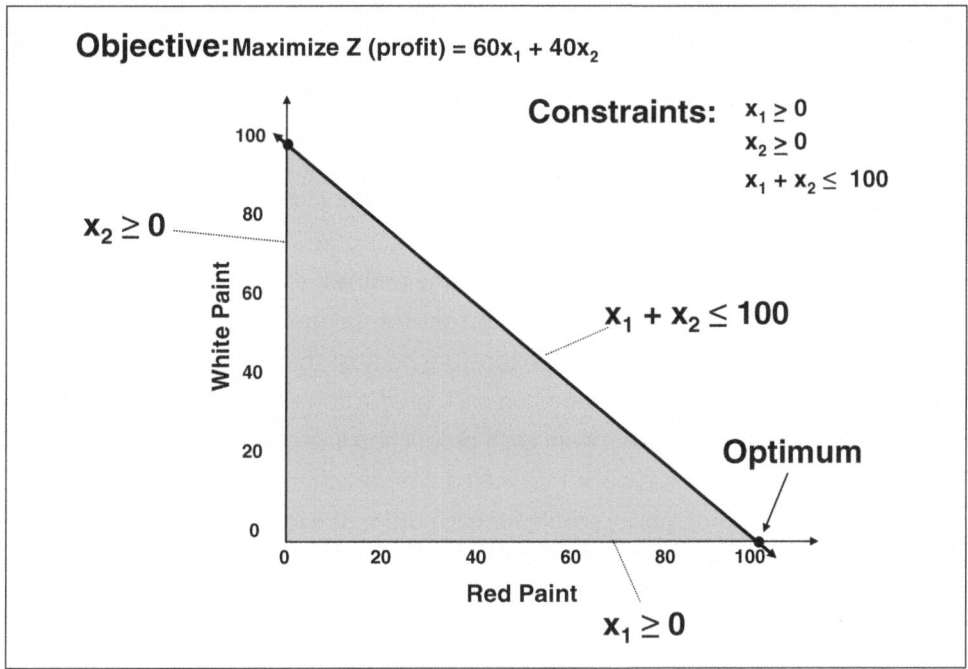

Objective: Maximize Z (profit) = $60x_1 + 40x_2$

Constraints: $x_1 \geq 0$
$x_2 \geq 0$
$x_1 + x_2 \leq 100$

Figure 7-1 Simple Optimization Example

If all manufacturing problems were as simple as this red and white paint problem, the solutions would be obvious and there would be no need to apply advanced mathematical techniques to solve them. As a second, slightly more complex example, consider a similar paint plant designed to produce paint on three different production lines in a single plant. It had been producing three different color paints; red, white, and blue, but the demand for blue paint decreased to the point that the manufacturer decided to only make red and white paint. Blue paint had been made on all three production lines, therefore eliminating the production of blue paint has made some production capacity available on each of the three lines. As a result of eliminating the manufacture of blue paint, line one has 10% available capacity, line two 6%, and line three 5%.

For this example, only white paint can be produced on line one and each drum of white paint produced consumes 2% of production capacity of that line. Line two can make either white or red paint, but a drum of red paint consumes 2% of production capacity, and a drum of white paint consumes 1% of capacity. Line 3 can only make red paint, and each drum of red paint consumes 2% of the available production capacity. The same profitability profile exists as with the first example, that is, a drum of red paint produces $60 profit and a drum of white paint produces $40 profit. Notice that although this is a fairly simple problem, the answer is not as obvious as it had been for the first example. Optimization techniques apply nicely here.

To organize the information for this problem in a manner that will help set up the optimization model, a simple table can be useful (Figure 7-2). This table displays all of the data in a manner that helps to clarify the problem to be solved. The three production lines are represented along the left side of the model, and the two products (red and white paint) along the top. The body of the table represents the amount of available capacity consumed on each production line for one drum of each product manufactured. The capacity in each production line available because the manufacture of blue paint has been terminated is represented along the right hand side of the table. The profitability profile for each drum of each product is shown along the bottom of the table.

product line	Consumed Capacity		available capacity
	Red Paint	White Paint	
Line 1	0	2	10
Line 2	2	1	6
Line 3	2	0	5
profit per drum Z	$60	$40	

Figure 7-2 Optimization Table for Example 2

The table is helpful for the construction of the objective and constraint functions. The objective function for this example is exactly the same as for the previous example.

Maximize $Z = 60x_1 + 40x_2$

Where:

Z is profit

x_1 is the number of drums of red paint made and

x_2 is the number of drums of white paint made

Adding complexity to this problem are the additional constraints. The five constraint functions for this problem are:

$2x_2 \leq 10$ (constraint on first production line)

$2x_1 + x_2 \leq 6$ (constraint on second production line)

$2x_1 \leq 5$ (constraint on third production line)

$x_1 \geq 0$ (cannot make less than 0 red paint)

$x_2 \geq 0$ (cannot make less than 0 white paint)

As with the previous example, the next step is to graph the constraint functions on the same set of axes. Figure 7-3 presents a simultaneous graphing of all the constraints with the shaded areas representing the potential solutions to the problem. As before, the optimal solution will be on the edge of the shaded region and at one of the vertices. The potential optimal points are highlighted on the graph by a bold dot. Adjacent to each of these dots are parentheses containing the ordered pair representing the coordinates of the point.

Plugging the values from the ordered pairs into the objective function $Z = 60 x_1 + 40 x_2$ yields the Z value for each of these points, which are displayed for the three most likely optimal points. The value of the point on the lower right of the shaded region (2.5, 0) has a Z value of 150. The value of the point just above it (2.5, 1) has a Z value of 190. And the value of the point on the upper right of the shaded region (0.5, 5) has a Z value of 230. From this analysis it is clear the optimal point is the point at the upper right of the shaded region (0.5, 5). This means the production mix on line 2 that maximizes the profitability per day of the plant involves using the available capacity to produce an additional 5 drums of white paint and an additional .5 drums of red paint. Notice that even though red paint is more profitable than white paint, the solution that actually results in the most overall profit involves making more white paint than red with the available capacity. This is because of the constraints on the process coupled with the consumed capacity for the production of each paint on each production line. This is most likely not the solution that would have been chosen by simple common sense.

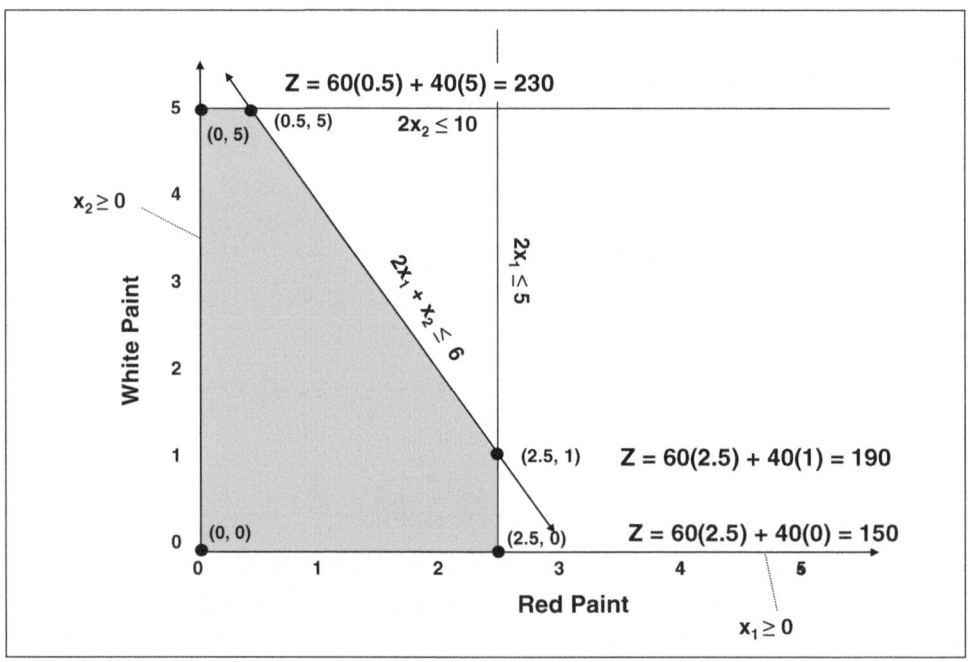

Figure 7-3 Graphical Representation of the Solution in Example 2

This simple example may help to demonstrate the basic concepts associated with optimization. Typically, real manufacturing problems will involve more variables and are much more complex. In the example, the objective function and each of the constraint functions are linear functions, which means that each equation graphs as a line and not a curve, making it a linear optimization problem. Linear optimization problems are the easiest to solve mathematically. The mathematical approach to solving a linear optimization problem is referred to as linear programming. Linear programming usually involves solving a number of linear equations simultaneously and an analysis of the vertices in a manner similar to what was done graphically in the example.

If either the objective function or any of the constraint functions is not linear, then you will need a different and more mathematically sophisticated approach to solve the problem. The mathematical approach most commonly used to solve a nonlinear optimization is referred to as nonlinear programming. Nonlinear programming involves much more complex mathematical approaches (such as gradient vector analysis) than does linear programming, and resolving a solution can take considerable computer resources. Other mathematical approaches such as neural network processing have also been successfully applied to optimization problems.

Applications of Optimization

Optimization software sees use in a number of different areas of manufacturing. One of the most common applications of optimization software may be to determine the optimal production schedule for a plant or a set of plants in a manner very similar to the example above. Optimization software can also be applied to process optimization, in which the optimal output may be the set point of a simple or complex controller (Figure 7-4). Other applications in manufacturing include optimizing a transportation route from the manufacturer to the customers and optimizing material and energy flow through a process.

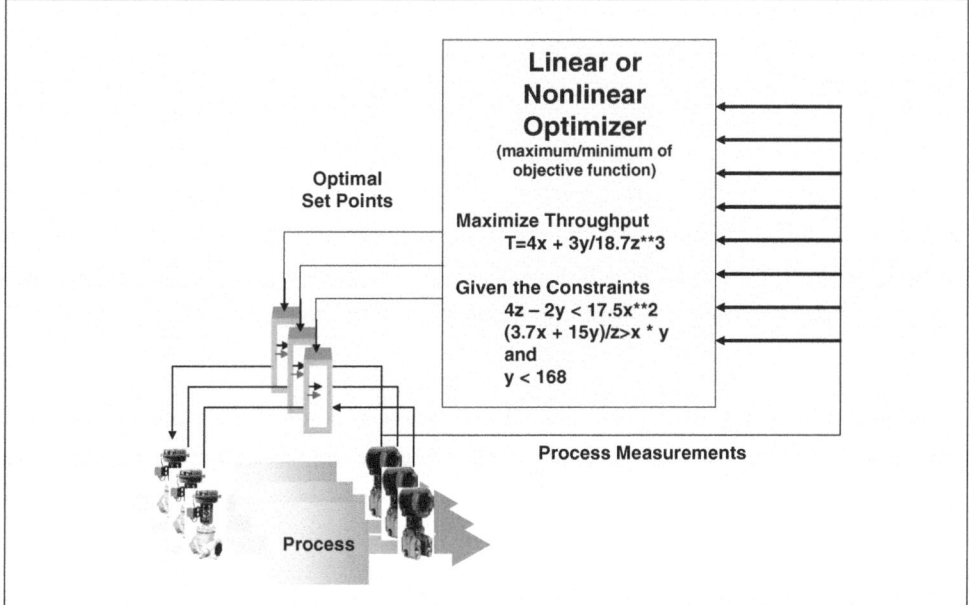

Figure 7-4 Optimization Systems

Optimization software can provide significant value in manufacturing operations, but there are a few shortcomings. The first is with classical linear and nonlinear programming, the models developed are static models while the processes being optimized may be dynamic. As the process dynamics change over time, the models may no longer match the reality in the plant, and the output may not be optimal. Second, mathematical optimization has traditionally been limited to a single objective function. Manufacturing operations involve multiple objectives, such as maximizing production at minimal cost to the business. For these problems, one of the two objectives has to be chosen as the objective function, and the other has to be relegated to being a constraint, which is suboptimal. Finally, optimization software requires talented people who know how to apply sophisticated optimization concepts and how to use the software. This talent is difficult to come by.

Constraint Analysis

One discipline becoming very popular is closely related to linear and nonlinear optimization. It is called constraint analysis. As you can see from the above examples, the optimal solution of any optimization problem is highly dependent on the constraints on the process. Often the most significant improvements can occur if you change one or more of the constraints. For example, in a manufacturing process one of the constraints may be the maximum possible flow of a liquid, which is limited by the size of the pump used to move the liquid. Certainly the process can be optimized within this constraint, but one may ask how the optimal solution value would change if a new pump goes in that can increase the flow. The optimizer can be reevaluated based on the new level of flow, and the value to the plant of the new pump can easily be discerned.

Significant improvement can result from analyzing the constraints around a process and determining which constraints limit the value of the objective function the most and whether or not you can change those constraints. Constraint analysis should always begin with the constraints that intersect at the current optimal solution. Those are the constraints having the most impact on operating the process in an optimal manner.

Optimization software is sophisticated and requires a high level of talent to apply effectively, but it can result in significant dividends when applied correctly. That is easy math.

Review Questions

1. What is an objective function in an optimization problem?

2. What is a constraint function in an optimization problem?

3. How many objective functions can be solved by mathematical optimization at any one time?

4. What are two common applications of optimization in industrial plants?

5. Specifically, what could be the difference between a linear optimization problem and a nonlinear optimization problem?

6. What is constraint analysis?

CHAPTER 8

Simulation and Modeling: A Look Before You Leap

After an explosion damaged their main spacecraft, the Apollo 13 astronauts were losing oxygen and power, and they had to rush to the lunar module to get that craft's system up and running. That procedure was less difficult to carry out because the crew had practiced it in simulators. Simulators give a look into a process or a procedure, which can prove invaluable when a real-life problem arises.

Modern automation systems operate manufacturing processes that produce products in an economical manner to meet market demands. Actions taken by the automation system have an impact on the efficiency and economics of the process and could even create a dangerous situation. Therefore, for training, testing, application development, and demonstrations, there is a need to simulate running an automation system in as realistic a manner as possible but in an environment that is safe and will not adversely impact an operating process. That is where process simulators come in. These simulators tie into automation systems and behave like a real manufacturing process.

Over the years, developers have created simulators using quite a few different approaches. Some of the earliest simulators, created before the advent of automation, were actually small, safe processes, such as the process cart in Figure 8-1. The basis of these simulators was the development of a set of physical equipment (such as flow, level, temperature and/or pressure measurement devices, dead time units, and units with a reasonable amount of capacity) designed to reproduce the characteristics of a real process. Engineers used these simulators, connected into a control system, to learn how to develop control strategies, tune loops, and monitor the control strategies as they operate. Operators could also use these simulators to learn how to use the control system to operate a process. Although this type of simulator did not match any specific manufacturing process, they were general-purpose enough to support training around the basic concepts of control and operation. Automation training centers still employ these physical process simulators today.

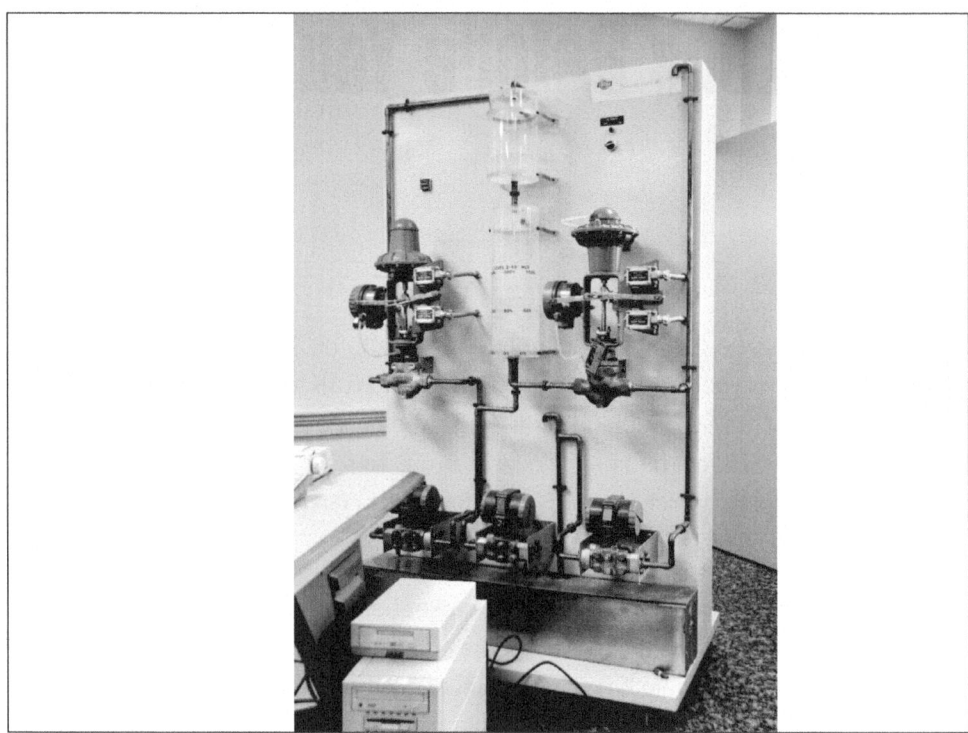

Figure 8-1 Physical Process Simulator

Another approach to simulation evolved along with the development of analog electronic control systems. Analog electronic control systems allowed for simple feedback control as well as more advanced control approaches such as feedforward control. Designed with these capabilities, electronic analog simulators were able to simulate all the functions required for these systems, including providing the gain, dead time, and capacity characteristics of manufacturing processes. The dead time and capacity (sometimes called lag) functions could typically be adjusted to match the capacity and dead time of any simple process. These simulators can be configured to provide a fairly good simulation of a simple process loop.

These analog simulators proved so useful that they were prepackaged, such as the one shown in Figure 8-2. Users can adjust the dead time and capacity for each loop via dials on the front of the simulation units to simulate process loops with vastly different dynamic characteristics. Analog electronic simulators were packaged in a manner that made them very portable, which increased their effectiveness for training. These "suitcase" simulators are still in use today.

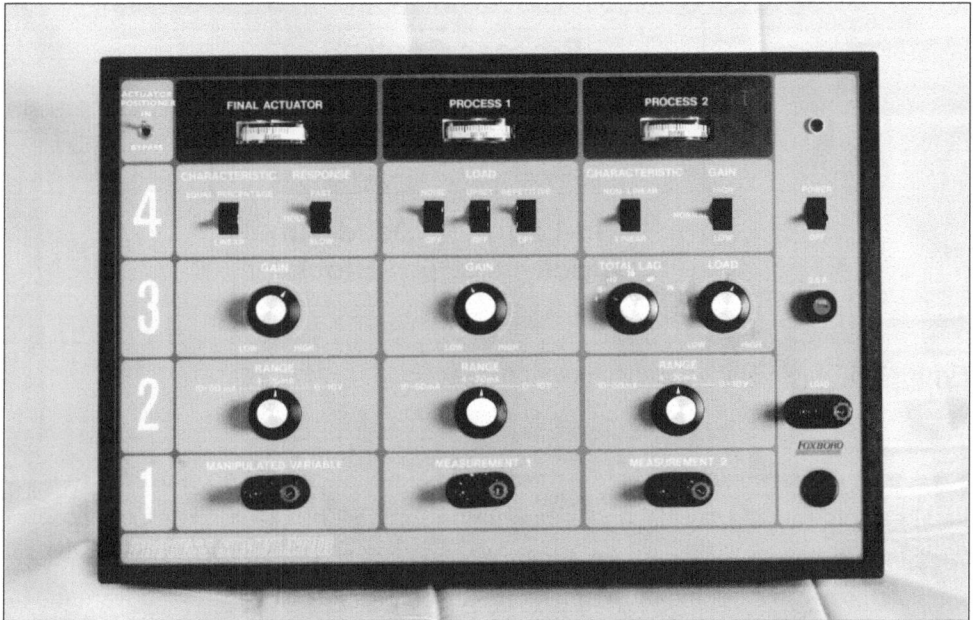

Figure 8-2 An Analog Simulator

With the introduction of digital computer-based control systems, the availability and sophistication of simulators increased significantly. The structure of a digital computer-based control system includes software blocks that essentially simulate the functionality of each component of the analog control systems that preceded them. This means they have capacity (lag) and dead-time blocks as part of their standard block set.

Unlike analog systems, using incremental blocks did not cost more money per component, but merely used computer capacity. This led the way to the development of a range of software block-based simulators, from simple single-loop simulators (Figure 8-3) to fairly sophisticated process unit-based simulators. These simulators often ran right in the computers in which the control strategies were also developed so they needed no external simulator. Many of these software block-based simulators have been out for 25 years or so and have been used in product demonstrations and training.

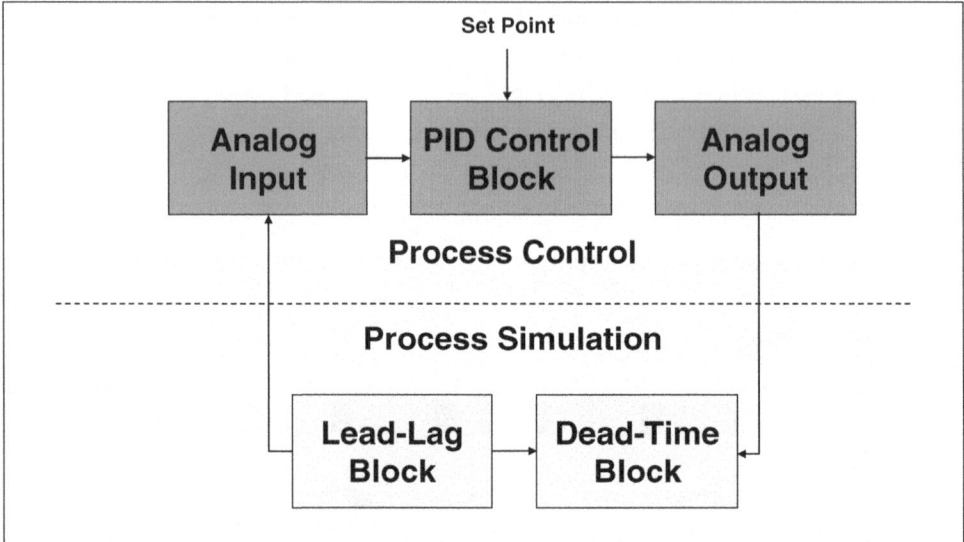

Figure 8-3 Simple Block-Based Simulation

Computer technology offered a much higher level of calculation capability than had been previously available, and although the software block-based simulators were fairly simple to build and use, the possibility of more sophisticated software simulations became apparent. Doing what they do best, developers took advantage of this calculation capability to create simulation software based on the use of "first principle" models.

First principle models are mathematical models of the basic physical, chemical, and biological relationships discovered by engineers and scientists over the centuries in their study of how things work. These models derive directly from the basic laws of physics, chemistry, and biology, which is why the experts call them first principle models. Thousands of first principle models have been discovered that mathematically describe what happens when different materials heat up; or different chemicals mix together; or material moves from one point to another; or other physical, chemical, or biological activities take place.

Digital computers offer a general computing capability ideal for simulating first principle relationships in software. To build a rigorous first principle model of a section of a manufacturing process, scientists define the process down to its basic operational components, identify the first principle relationships that define what takes place in each component, and develop software models of those components based on the first principle relationships. By doing this, very accurate and rigorous simulations of specific manufacturing processes can appear in software, and behave very much as the manufacturing processes would behave in ideal conditions.

Building these models requires a significant degree of scientific and software expertise, and a number of simulation companies specializing in this level of intellectual property came into existence through the 1980s and 1990s. In the nuclear power industry, for example, quite a few operator training control rooms, completely driven by a rigorous simulation of the plant, behave exactly like the plant's actual operating control room. This provides an ideal environment for training the operators in not only how to run the plant, but also how to respond to dangerous situations. This type of training is also invaluable in other industrial plants.

Software-based first principle models are ideal for the development of rigorous process simulators, but these models can see use in a number of other ways that can be very helpful in industrial operations (Figure 8-4). Designing manufacturing processes that perform in the expected manner has been an engineering challenge for decades. First principle model software can be very useful in designing individual processes and entire manufacturing plants in a manner that ensures the expected level of performance when the plant is up and running. Plant design software exists that is very specific to the type of plant being developed.

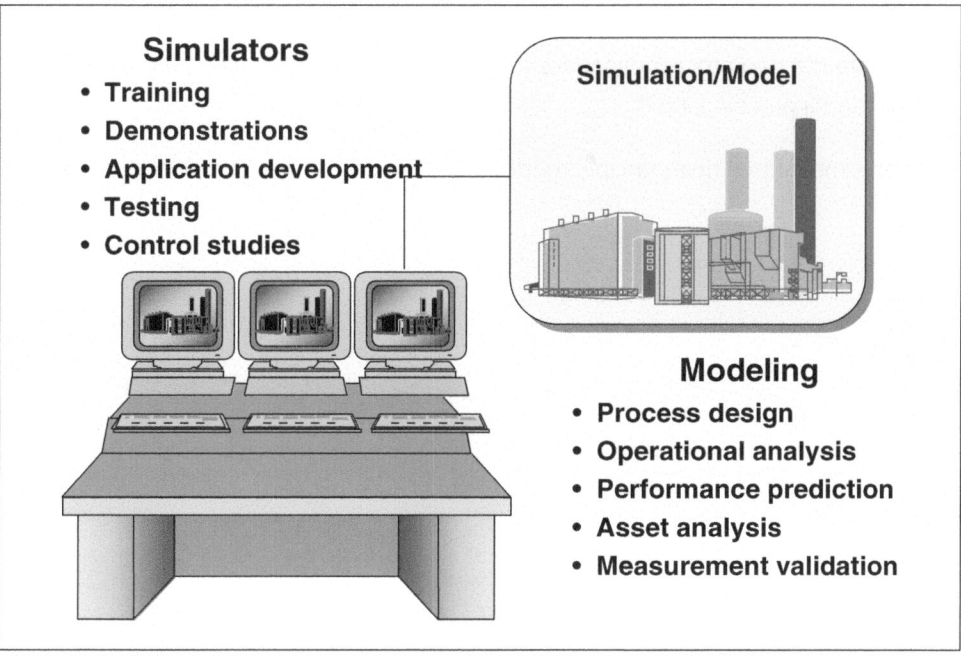

Figure 8-4 Software Simulation and Modeling

One interesting side benefit of designing a plant using this kind of design software is that once the plant is ready to go, a rigorous model of the plant that defines how the plant should work in ideal conditions also exists as an output from the modeling software. A model of this type can run and then be compared to how the actual plant is operating to try to identify areas of the plant that need attention. It can also see use in other ways within an operating plant, such as simulation, performance prediction, asset analysis, and measurement validation.

New applications of rigorous first principle software models are appearing all the time. Although the development of these models takes significant effort and expertise, the benefits in terms of increased efficiency and safety can justify the expenditure with payback in a very short time. As industrial operations strive for improved business performance, simulation software will have to keep growing.

Review Questions

1. What are four different uses for simulators in the process industries?

2. What are three different technological approaches to the development of simulators for the process industries?

3. What are the two process dynamics that simulators must be able to imitate while simulating a process loop?

4. What is meant by "first principle models" when used in an engineering context?

CHAPTER 9

Safety Management Systems: Expect the Unexpected

Not to sound alarmist, but dangerous situations that could lead to equipment damage, injury, damage to facilities, or even death are an everyday possibility in process plants. That is not hype to sell a book; that is a simple reality.

Process manufacturing often involves volatile fluids, explosive materials, high temperatures, high pressures, dangerous chemical reactions, poisonous chemicals, and other potentially lethal materials and conditions. As long as the production process is working normally, the potential for events leading to dangerous situations, such as the overheating of a pressurized vessel causing an explosion in the vessel, is usually very manageable by operators working through the automation system. But unexpected things, such as equipment failures and operator errors, can and do happen in complex production plants. When unexpected incidents occur, corrective action needs to happen quickly enough to prevent these potentially dangerous events or minimize their impact.

The topic of safety response has already been addressed to a limited extent in this book. Electronically controlled pneumatic valves have been designed to either open (air-to-open) or closed (air-to-close) on the increase of the signal to the valve. This means with the unexpected loss of power to the plant, each valve will move to either the open or closed state depending on the valve type. Engineers designing plant piping and instrumentation select the type of valve for each location in the process. If this is done correctly, each valve in the plant will automatically and immediately move to a predetermined safe state upon loss of power to the plant. Although this is a rudimentary aspect of safety management, it can be a very effective loss prevention approach that covers a single event—a plant power loss.

Another example previously discussed was the exception logic of a batch control system. This exception logic software runs in parallel with the normal operations of the plant. The software monitors the operating process to identify exception events that can lead to unexpected and even dangerous situations. Once it detects and identifies an event, the software determines and executes the most reasonable response. Exception logic in batch automation software operates almost like an internal safety management system.

In continuous process plants, such as oil refineries and petrochemical plants, the equivalent of exception logic is not built into the process control software. Part of the reason for this is that to be effective, a safety system for a huge, continuously operating process must continually monitor all aspects of the process and respond very, very quickly once it identifies a potential event. It would be difficult to get the necessary speed of response by putting the safety management software into the same software system as the process control software.

Second, one of the events that may present the most danger in a large continuous process plant is the failure of the automation system. With the failure of the automation system, not only is all process control lost, but the operator's window to the process is also lost. The plant is essentially running blind and perhaps even out of control. If the safety management system were part of the process control system, upon system failure both would be lost to the plant. This would present an unacceptable risk.

To deal with the possibility of a dangerous event occurring in a continuous process plant, a new class of computer-based automation system began to emerge in the 1980s. This new class of system, called safety shutdown systems, or just safety systems, was designed to go into continuous process plants along with, but independent of, the distributed control systems (DCSs) in the plants (Figure 9-1). The DCSs would operate the plants under normal circumstances, basically unaffected by the existence of the safety system in most instances. The safety systems connected through instruments and to valves and other actuating devices, and continually monitored the process to detect predefined dangerous conditions. If a dangerous condition was identified, the plant would immediately cycle through a predefined safe shutdown sequence and come to a complete stop, thus preventing the dangerous condition from becoming a catastrophic event.

Speed of detection and response to predefined dangerous conditions is, and was, a primary design requirement for safety systems. Since programmable logic controllers (PLCs) are designed with almost the same speed characteristics, the first safety systems emanated from PLCs.

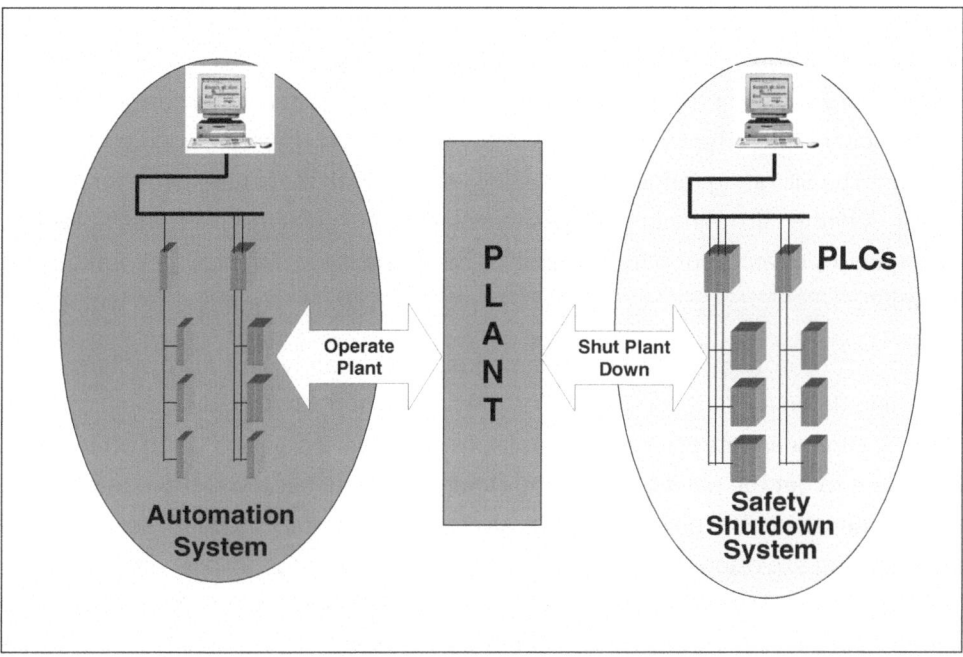

Figure 9-1 Safety Shutdown System

One important consideration with respect to the design of a safety system is that its failure could lead to dire consequences. Safety systems typically do not see action very often, but when they are needed they must perform. With this in mind, safety systems must be continuously available to the process through the redundancy of modules and communication networks. With a redundant system, if a primary module fails, a redundant module takes over to perform the functions required. Two approaches to redundancy see use in safety systems: dual module redundancy (DMR), in which each active module has a duplicate redundant (backup) module, or triple module redundancy (TMR), in which each active module has two active backup modules. With TMR, if any two modules fail, the third will still perform the necessary functions. There has been an active debate in industry over the past two decades with respect to the relative merits of DMR and TMR systems. The net tradeoff appears to be that DMR systems are typically less expensive, but TMR systems provide a higher level of availability, i.e., they are more reliable. There are other differences, but cost and reliability appear to be the primary determining factors.

Standards bodies, such as TÜV (an industrial-company-sponsored standards organization that started in Germany), IEC (International Electrotechnical Commission), and ISA have attempted to provide a means to measure the integrity of safety systems and components by rating the safety integrity levels (SIL) of various components so industrial users can make intelligent decisions as to the most appropriate way to approach the selection of the safety systems and components for their operations. The SIL is the relative level of risk reduction provided by a safety function. The SIL rating ranges from the lowest rating of SIL1 to the highest rating of SIL4. A higher SIL rating implies greater risk reduction. The risk level and potential consequences of an event determine which SIL rating is most appropriate for a plant operation.

One of the downsides to the implementation of a safety shutdown system is shutting down a plant can be very costly. For large refineries or petrochemical plants, it can mean lost production at millions of dollars a day, and it can often take a few days to get the plant back up and running after a shutdown. If the safety system shuts down the plant in response to a true dangerous situation, such as an impending explosion, then the cost is not much of an issue. But as with any programmed system, safety systems sometimes make mistakes and shut down plants when it is not warranted. Unwarranted shutdowns do occur and can be very expensive mistakes. Also, shutting down and restarting industrial processes tend to be the most dangerous and stressful phases of operation for the equipment and the plant.

As an answer to this, a number of process manufacturers, such as the Dow Chemical Company, have been advocating that shutting the plant down may not be the only possible response to the detection of an impending event. Going to a reduced level of production, without shutting the plant down completely, may be just as effective a response.

Considerable research has been undertaken by process manufacturers and safety system suppliers alike to try to understand the type of response that may be most appropriate to any detected impending event. The result has been that a major transition has taken place over the past few years from the uncompromising principles of safety shutdown systems, in which shutting the plant down was considered to be the only responsible action, to a more open approach under the banner of safety instrumented systems (SIS), which bring a process to a predefined safe state upon the detection of an impending event (Figure 9-2). There are certainly unexpected circumstances for which the most appropriate response is to shut the plant down. But there are many other circumstances that may require only shutting down of a section of a plant or even just slowing down production to some predefined safe level until the causes can be dealt with.

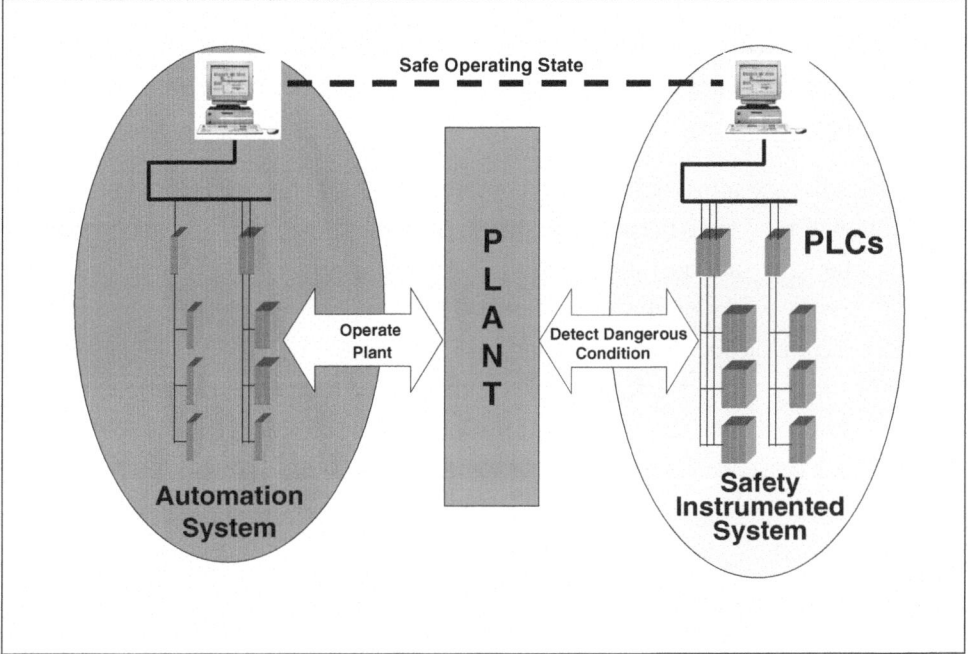

Figure 9-2 Safety Instrumented Systems

In the case of a SIS implemented to execute a safe response to impending events other than just shutting down the plant, considerably more effort has to go into the appropriate response (or responses) by the SIS and to programming each response. Now we start to get into a cost issue for the executives to ponder. While the cost of implementing a SIS that has variable responses can be significantly more than that of a safety shutdown system that simply drives the plant to a shutdown state, the avoided cost of lost production provided by an SIS can typically cover the incremental engineering cost of the SIS in very short order. Looking at it another way, a single avoided unnecessary shutdown can pay for considerable engineering time.

Coordination between the process control system operating the plant and the SIS needs to be considerably greater than with a shutdown system. Upon the detection of a problem that requires a slowdown of production, the process control system has to be told how to control the process by taking input signals from the SIS that let it know the new set point setting(s) required to operate the plant at the new level.

Standards bodies such as TÜV and ISA have traditionally insisted on the separation of safety and automation systems. They are trying to ensure that a failure in one will not adversely impact the operation of the other. There are typically very sound reasons for the rules that come out of these organizations, such as OSHA mandates, and both the reasons and the rules should be very well understood prior to the implementation of automation and safety systems.

Today, safety systems (both safety shutdown systems and SISs) are separate from plant automation systems because the standards bodies deem the separation to be in the best interest of safe operations and the best way to protect people and the plant. While it appears as though this separation will be the rule for years to come, more and more communication between safety and automation systems is taking place due to SISs. At some point in the future, DCSs and safety systems may converge to form a new class of combined safety and automation systems. But with the risk that is at stake, this will most likely not happen for some time to come.

Review Questions

1. Why are independent safety systems required in process plants?

2. What is a safety shutdown system?

3. What is a safety instrumented system?

4. What are two standards organizations involved with developing standards around safety systems?

5. The safety integrity level ratings of safety systems are designed to convey what information?

CHAPTER 10

Automation System Security: Checkmate

A report hits the airwaves: A foreign hacker who penetrated security at a water filtering plant near Harrisburg, PA, is under investigation by the FBI for planting malicious software capable of affecting the plant's water treatment operations.

The hacker tried to covertly use the computer system as its own distribution system for e-mails or pirated software, officials said.

"The concern was high because it is a computer that controls an important infrastructure system, and if for some reason it caused it to fail, it would have disrupted service," an FBI spokesman said.

The hacker, operating on the Internet, tapped into an employee's laptop and then used the employee's remote access as the point of entry and installed a virus and spyware in the water plant computer system. Following the intrusion, the plant changed all passwords to the system and eliminated remote access to the system.

"This is very common; computer hackers try to gain control of systems to use them as a resource to distribute e-mails, pirated software. It does not appear that this particular computer was hacked into for any other reason," the FBI spokesman said.

This incident really happened, and this was not a rare occurrence.

As we learned from previous chapters, unplanned downtime is unacceptable and just kills profits. Now think for a moment about what could happen if some person or some program sneaks into a manufacturer's system through the Internet and pushes a few buttons here and a few buttons there. At best a system could go off line. At worst, it could be a catastrophe.

Automation system security is there to minimize or eliminate any probability of a disruption to the process automation system. When we talk about system security, it is not just a piece of hardware and some software. Rather, it is a unified mindset the entire company shares to keep the bad guys out.

It can be easy to confuse automation system security with the safety management systems discussed in Chapter 9. While these two topics do overlap to some degree, they have very different focuses. Safety management systems detect potential or impending unsafe events in the production process and respond to those impending events in a manner that ensures safe operation of the plant. Automation system security, on the other hand, prevents any actions originating in the automation system or passing through the automation system from having a negative impact on the production process. Clearly there could be actions originating in the automation system that might lead to unsafe conditions in the process, causing the safety management system to respond, but automation system security is designed to stop such actions from happening in the first place. Safety management systems and automation system security are very important to the safe operation of an industrial plant.

With the introduction of distributed control systems (DCSs), new concepts associated with the security of the process in a distributed environment were developed. Since the most critical DCS function with respect to process security is the control of the process, suppliers introduced a number of approaches to ensure secure control. The most obvious one was the development of redundant controllers, so if one of the controllers had a problem, the other could take over control. Significant effort went into the design of the software associated with redundant controllers to ensure the process could tolerate faults occurring in one of the controllers. As has been discussed, for critical control functions, triple redundancy could be implemented, often leading to a predicted mean time between functional failures in the hundreds of years.

System management software was also developed for DCSs to improve the level of security of the systems. System management software controlled secure communication across the DCS networks, provided loading of control software, conducted system-wide failure analysis and management, and provided many additional security-based functions. With the advent of these techniques, there was a significant reduction in the probability of a failure in the DCS causing a process problem.

A whole new set of security issues surfaced as DCSs became more open. Prior to the late 1980s, most DCSs were based on proprietary operating systems and networking architectures. Because of this, DCSs were fairly isolated and were seldom connected to corporate networks, making intrusion into DCSs from the outside world highly improbable. As more commonly used and standard networks and operating systems were incorporated into DCS architectures, the tendency was to connect the DCSs into corporate networks in order to make the wealth of data contained in the automation systems available to the business systems.

As this took place a new security threat arose: the possibility of a communication from the corporate network impacting the operation of the automation systems and in turn impacting the process. New security measures were required to protect the process from dangerous intrusions through the now-connected cyberspace. Today, when most automation specialists discuss automation system security, they are typically referring to the challenges associated with cyber security.

Cyber security is an enormous field of study, in flux and with much work underway to improve security approaches and methods. There are many challenges facing the appropriate implementation of cyber security. One of the largest challenges is balancing two primary objectives in most industrial organizations: 1) the need for communication flow between automation and corporate information systems and 2) the need to protect the automation systems from unwanted and unexpected messages from the corporate systems. These two objectives are almost diametrically opposed. On the one hand, corporate IT organizations must provide easy-to-use, open communications, and on the other hand they need to ensure that no communications are made to the automation systems that may have potentially damaging consequences to the already dangerous production processes.

This issue was not too difficult to manage when the corporate systems were exclusively contained within the walls of the corporation, but with the rise of open communication networks, such as the Internet, the potential for problems increased exponentially (Figure 10-1).

A number of software and hardware tools have been deployed over the years to try to address cyber security issues. Software firewalls, designed to detect abnormal messages and stop the throughput of potentially dangerous messages, have been implemented between the corporate network and outside networks and between the corporate network and the automation systems. Both approaches have been successful but as Internet hackers have increased their activities, the firewalls have had to be continually expanded to be effective. Also, firewalls can, at times, make it difficult for desired communications to pass through the different levels of the systems architecture.

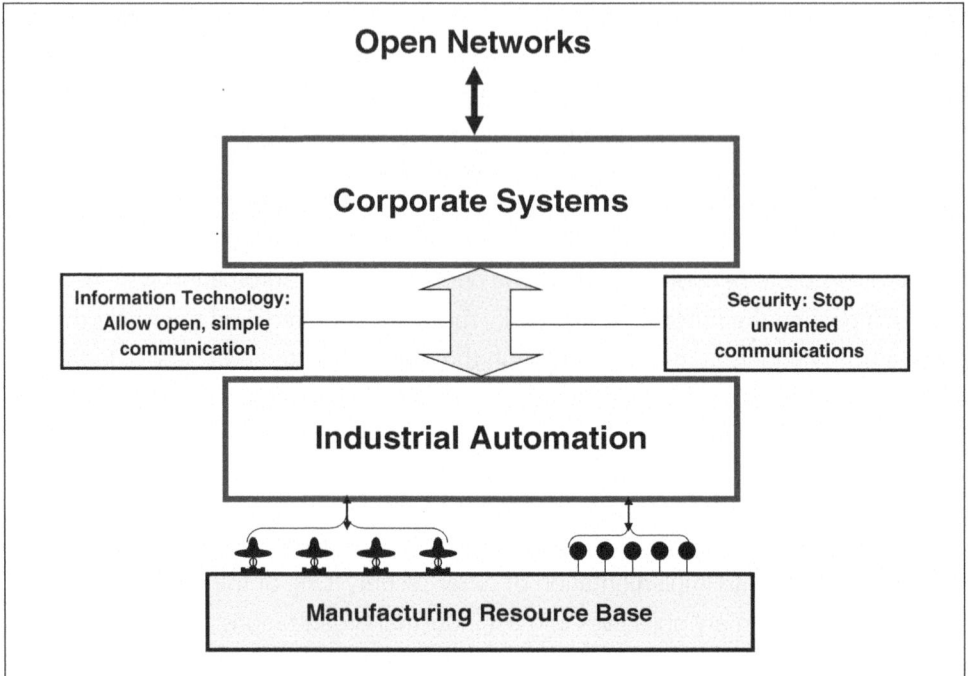

Figure 10-1 The Cyber Security Challenge

Other software approaches such as message encryption and data validation have had some success. Passwords and other user authentication techniques have been employed to ensure that no unauthorized user is accessing either the corporate network or the automation systems. Virus detection software has been developed and employed to ensure destructive viruses do not get access to the systems. Many other approaches have been used to ensure the security of the automation systems and the security of the production process are maintained at as high a level as possible.

Unfortunately, once automation systems open up to corporate networks and to the outside world, there is no foolproof security approach that will guarantee anyone will be free of hackers. The real issue is how industrial companies should go about approaching the security of the processes and the automation systems in the most effective manner.

It would be nice if there were a simple set of technologies a manufacturer could integrate that would provide a secure environment, but this is just not the case. The reason for this is that every industrial company has its unique set of requirements in terms of communications and security. The approach to automation systems security must therefore be developed to meet the exact needs of each enterprise. By the way, if there were a cookie-cutter approach to security taken by all industrial companies, it would become much easier for hackers to learn the approach and work around it, which would ultimately defeat the purpose of the security system.

Even though there is no cookie-cutter approach to the implementation of security tools, there is a general set of critical success factors that have proven to be very effective in the design of automation security systems. The critical success factors are:

1.   Take a long-term view of security

2.   Assemble the correct team

3.   Partner with companies competent in automation security

4.   Build on the foundation of what is already in place

5.   Design security solutions that fit the requirements

6.   Be vigilant

7.   Remember, security is a mindset, not just a solution

Taking the long-term view of security is an essential factor because plants and automation systems technologies tend to have extended lifecycles. Automation systems, unlike their business counterparts, have effective lifecycles measured in decades. With this in mind it is essential that a corporate security program has senior management sponsorship and the concepts and practices of security become a common way of doing business for all employees.

To be most effective, the corporate security team has to have representation from a number of different functional organizations within the company, including operations, information technology, business leaders, and engineering. The members of the security team should be selected based on the skills and experience they bring to the issue of security rather than their availability. Some of the company's top professionals should be part of the security team. Security is a serious business and must be treated as such by the corporation to be effective and to be taken seriously throughout the organization.

There are vendor companies in the marketplace that have developed significant intellectual property in the area of cyber security. The technologies supporting cyber security are continually advancing, and it is difficult for any industrial corporation to stay on top of all the advancements. Companies with specialized offerings in cyber security make it their business to understand the application of the technologies as well as to keep up with changes in technology.

Building on a solid foundation of what is already in place may seem to be an obvious success factor, but it is one that is sometimes underplayed. Every manufacturer most likely has a rich collection of technologies of various types, such as business systems, DCSs, PLCs, supervisory control and data acquisition (SCADA) systems and the like, already in place throughout their operations. These systems will most likely not be replaced or upgraded as part of a cyber security program; the cost is typically too high. Therefore, a complete assessment of what is installed and the expected lifecycle of each installed system is an important starting point.

These installed systems must be analyzed from the perspective of the external access points into each system, the data transfer requirements, the probability of an intrusion across each transfer point, the data storage and management approaches deployed, and the existing security approaches in place. This analysis will provide a good assessment of the current state of security and of potential security issues.

As has been discussed, one size does not fit all industrial operations when it comes to cyber security. Today every effective cyber security system has to be customized to the exact needs of the industrial operation. Developing a design that exactly fits the requirements of each industrial company is essential and can be effectively accomplished. Doing this can be a costly undertaking, but the alternative can be much more costly in terms of economics and the health and safety of plant employees.

Once an industrial operation agrees upon a cyber security system and approach, it doesn't just end there. It is not like you can just forget about the system once you implement it. Rather, the cyber security team has to keep on top of the system and continuously update it. After all, hackers are learning new ways to infiltrate systems every day. It is like an ongoing chess match; as hackers move to find new ways to slam your system, you need to have a team that will stay one step ahead. Checkmate.

Review Questions

1. What is the fundamental difference between safety management systems and automation system security?

2. What is the definition of automation system security?

3. How did the introduction of open computer networks, such as the Internet, impact automation system security?

4. What are the primary tradeoffs when implementing an automation system security system in situations in which it is desirable to have communications between the automation systems and other systems in an industrial company?

5. Why is it important to avoid a cookie-cutter approach to automation systems security?

CHAPTER 11

SCADA Systems: Beyond Four Walls

Automation systems such as distributed control systems (DCSs) and programmable logic controllers (PLCs) operate within plants and factories. However, there are quite a few industrial operations that require a level of monitoring and control that extends beyond a plant.

For example, in the production of oil and gas there are pipelines that transport the raw and finished products over great distances. These pipelines require a level of monitoring and controls similar to that provided inside plants, but they also have unique requirements and characteristics. This is also the case for water distribution systems and for power transmission and distribution systems.

Automation systems designed to manage the monitoring and control of these types of transportation and distribution systems over distance are supervisory control and data acquisition (SCADA) systems (Figure 11-1).

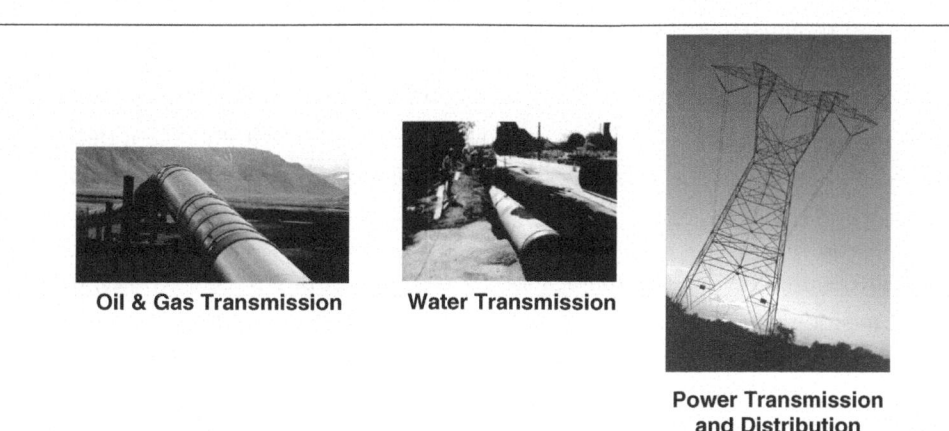

Oil & Gas Transmission   **Water Transmission**

Power Transmission and Distribution

Figure 11-1 SCADA System Applications

Before taking a closer took at SCADA systems, it is worth noting the acronym SCADA has actually assumed a second meaning, which has caused significant confusion. As personal computers began to provide some of the functions traditionally found in DCSs, such as operator interfacing and process control, some of the companies providing the software that operates on PCs started referring to this software as SCADA software. When this happened, SCADA literally had two very different meanings. It is somewhat like "lead" and "lead." One means a position at the front. The other hurts when you drop a big chunk of it on your foot. One use of "SCADA" was then monitoring and controlling over large distances, and the other use was PC software for industrial automation. Automation professionals often use the acronym SCADA without clarifying which meaning they are talking about. To this day, SCADA still has both meanings, so if the meaning is not obvious from the context of the discussion, then it is very important to ask. For this chapter, SCADA will refer to monitoring and control over distance.

As the requirement for automation capability over large distances arose, it became obvious that the primary enabler for these systems would be efficient and cost-effective distance communications. The land-line telephone systems first in use were expensive to build and maintain and could only transmit limited amounts of information. The science of telemetry, originally developed for rockets and weather balloons, was based on distance communications devices such as radio, microwave, or satellite transmitters. Applying this technology to two-way communication—measurements and control instructions—tremendously expanded the capabilities of SCADA systems. (Perhaps somewhat confusingly, "telemetry" also came to be applied to two-way wireless communication). However, such communication systems were expensive, so users tended to place a number of instruments in close proximity. This meant a single distance communication device could be used to keep the cost of the overall system down.

Since many of these measurement and control devices were in remote areas along pipelines or transmission lines, powering them became a significant design (and cost) issue. If no local power source was available, power sources such as batteries and solar cells came into play (Figure 11-2).

Just as with other telecommunications and computer-based technologies, the cost of telemetry has declined significantly over the past few decades. As telemetry technology evolved and the price of distance communications devices fell, the economic viability of these systems increased significantly, and the proliferation of SCADA systems likewise increased.

Figure 11-2 Sample Telemetry Devices (some with solar panels)

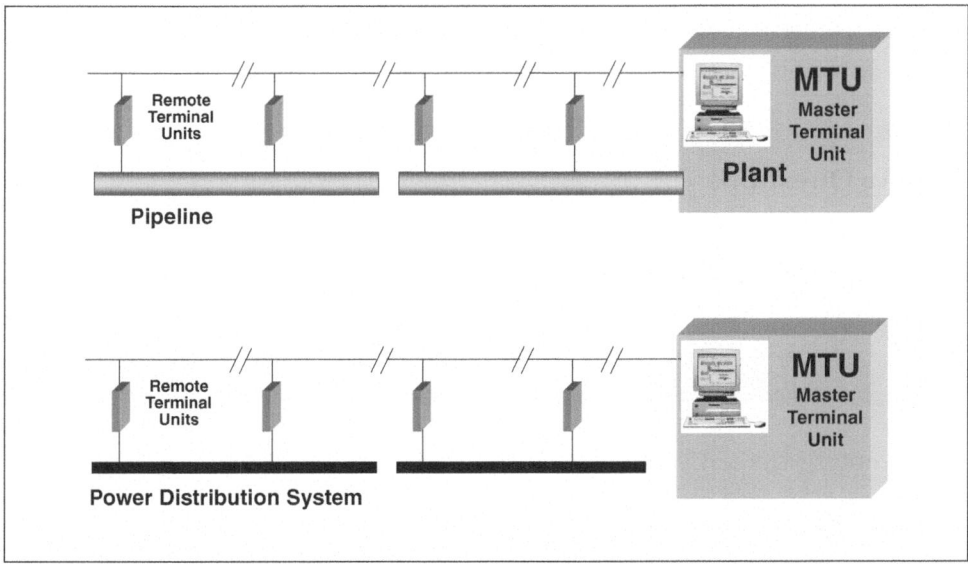

Figure 11-3 SCADA Systems

The architecture of a basic SCADA system is a centralized monitoring location at which a master computer system called a master terminal unit (MTU) connects by distance communications networks to remote devices or computer systems referred to as remote terminal units (RTU). The MTU provides the overall supervision and communication management of the SCADA system and the RTUs provide the remote measurement and control functions (Figure 11-3).

In early industrial SCADA systems, the local devices would typically be unique to the application or industry the SCADA system was being installed into, and therefore the interfaces between the RTUs and the devices would be different for water, power, and oil and gas. The software in the RTUs and the MTUs would typically also be unique to the industry, although some general-purpose automation software was used. Because of the specificity of devices and software for SCADA systems, vendor companies in the early going focused their systems on one industry. There were SCADA companies strictly focused on oil and gas, others on water, and others on power transmission and distribution.

Over the past decade, as telecommunications devices have become more abundant and available at much lower cost, much of the initial challenge associated with the development of SCADA systems has gone away. That means DCSs, PC-based automation systems, PLCs, or a combination of these technologies coupled with low cost but fairly reliable telecommunications devices have taken over the traditional SCADA applications. This has required the development of SCADA software that operates in these systems. It appears as though the age of independent, incompatible SCADA systems is moving toward a more unified automation approach.

Review Questions

1. What phrase does SCADA stand for?

2. What is telemetry?

3. What are three early industry applications that required SCADA systems?

4. The controllers in a SCADA system are referred to by what name?

5. The coordinating computer in a SCADA system is referred to by what name?

6. What is another use of the acronym SCADA in industrial automation?

CHAPTER 12

Quality Management: A Tale of Two Processes

Quality management is truly at a fork in the road for discrete manufacturing and process operations. The idea of quality management has evolved very differently in discrete manufacturing than in process operations. Traditionally, in discrete and process operations, the primary approach to quality management was after-the-fact quality analysis and control. That is, once a plant finished making the products, the manufacturer checked them to determine if they had come out properly. In the case of discrete manufacturing, this involved visual inspection and perhaps functional testing. In process manufacturing, where the products were typically liquid or gaseous, visual inspection served little purpose. Instead, the products made in process operations went to laboratories for testing to determine quality.

Today, as it was traditionally, the response to off-spec quality in process and discrete operations is also quite different. In process operations, the manufacturer may be able to mix off-spec product with other product or rework it in the process to correct the problem. In discrete operations it is often very difficult to repair the problem in finished products. A glaring example would be if a discrete manufacturing operation produces door hinges, one of the production operations may be to drill holes in the hinge plate for attaching it to the doorframe. Once the manufacturer drills the holes, if they are not in the correct position it is very difficult to correct the problem. The holes already exist and the part is defective. You can hear the *cha-ching* of a defective part hitting the trash bin. Therefore, the consequences of poor quality in discrete manufacturing operations can be more severe than in process plants.

Another major difference between discrete and process manufacturing is process control systems are inherently quality management systems. Process control systems tend to directly measure and control the quality variables in the process, while discrete manufacturing control systems typically do not. In discrete manufacturing operations, the quality variables typically do not undergo measurement until after the manufacturing operation is complete merely because the real-time measurement technology for the quality variables does not exist. With the drilling operation in the door hinge plant, the diameter of the hole is a critical quality variable. But it is difficult to impossible to measure the diameter while the hole is being drilled. Therefore, it undergoes measurement after the hole exists, and if the hole is too large or not round, the part is a throw-away. Once again, *cha-ching*. On the other hand, in producing gasoline, one of the key quality variables is the percentage of octane mixed into the gasoline by volume. Since the octane percentage is directly controlled by the control system, the control system is controlling not only the making of the gasoline, but also the quality.

A second difference between managing quality in process and discrete manufacturing is typically a defect in a part of a discrete manufacturing operation causes the manufacturer to discard the part. A variance from the desired value in a process operation can often be remedied by correcting the process to blend in a mixture that returns the value to the desired quality level.

These differences between discrete and process manufacturing were not very significant from an overall quality management perspective until the science of quality management started to develop in sophistication after the 1940s. During the war years significant manufacturing operation advancements came into play. Walter A. Shewhart championed quite a few of these advances from his labs at Western Electric Corporation. Later on two of his protégés at Western Electric, W. Edwards Deming and Joseph Juran, expanded upon Shewhart's ideas. Deming and Juran often earned praise for the quality movement that emerged over the following three decades.

It is important to realize the manufacturing operations that most concerned these three quality pioneers focused on discrete manufacturing. Since in most discrete operations, manufacturing defects are the critical quality control issue, the focus was on defect reduction. Since it was impractical to measure the quality variables of every part made, the manufacturers had to develop their approaches based on sample parts. When a defect was detected in a sample, it typically was not very clear how many other parts contained the same defect or even what part of the manufacturing process was the root cause of the defect.

The new field of statistical analysis addressed the problem. Manufacturers randomly selected and inspected parts and they developed a statistical profile of the defects.

Statistical tools such as Shewhart charts and X-bar and R Charts were able to help in the analysis of defects and in determining the root causes of defective parts. Manufacturers were then able to correct the root causes and implement processes to ensure they would not occur in the future. This is where the phrase "continuous improvement," which typically refers to the continuous reduction in defects in manufacturing lines, came into play. Programs like 6σ (Six Sigma) also showed extreme success in the reduction of defects. The term 6σ implies the variability from specification in the manufacturing operation is so little that only parts of the six standard deviations from the specification are defective. This is a very small number of defects.

- **Defects per million parts made**
- **Continuous improvement**
  - **Continuous defect reduction**
- **Cannot measure as product is made**
- **Statistical Process Control**
- **Discrete statistics for analysis**
  - **Identify and correct root cause**
- **Six Sigma–limit on defects**

Figure 12-1 Quality Management in Discrete Manufacturing

The sciences of statistical quality control, statistical process control, and total quality management all had their genesis in this movement.

Keep in mind this movement was almost completely focused on discrete manufacturing operations (Figure 12-1). All of the language and statistical analyses relied on a parts-based operation. In contrast, how would you define "defect" in the production of gasoline from crude oil? In these process operations the quality control system does not wait until after the product is made to measure and control the critical quality variables. As a part of the manufacturing operation, the critical quality variables constantly undergo measurement and control. This does not mean there are not quality problems in process manufacturing. There certainly are. There are some quality variables in process plants that are not directly measured either due to the cost or availability of the measurement devices. For these variables, the process of quality control tends to be much closer to that of discrete manufacturing; although technologies, such as analytical measurements, inferential measurements, and factor analysis have provided opportunities to treat these variables as though they are being directly measured. Samples of the manufactured products are sent to either onsite or offsite laboratories for analysis. If they are off spec, the control system undergoes an adjustment and the manufacturer either mixes or reworks the finished product back to spec. Sometimes the product is unrecoverable and they have to toss it away.

Another interesting and important difference between process manufacturing and discrete manufacturing that impacts the approach to quality management is process manufacturing is much more capital intensive, with far fewer process operators than discrete manufacturing. The major emphasis of most continuous improvement approaches coming out of the quality movement is on team-based improvement activities. This team-based approach tends to have much more effect in operations in which there are larger numbers of people working to the same end— discrete manufacturing operations. If you want to look at the team-based approach, check out Toyota Motor Corp. Their approach is legendary.

In process plants it is not unusual to only have one or two operators overseeing large segments of a plant. If there is more than one, they most likely are a very highly coordinated team already. When Joseph Juran developed his models for Total Quality Management, he often proposed starting with quality teams and moving to making continuous improvement part of every person's daily work. In process operations it may be more appropriate to go right to individual continuous improvement activities instead of the initial team-based approach used in discrete operations.

The point is that quality management in process manufacturing environments requires a different mindset than in discrete manufacturing operations. As statistical process control (SPC) started to gain popularity in discrete manufacturing operations during the 1980s, a number of process manufacturing operations purchased SPC packages and tried to apply them to their process operations in response to market hype. Many installed the software to see what value it might provide. Not surprisingly, the results were less than spectacular. Discrete statistical analysis tools do not provide the same information when applied to continuous variables. Statistical control, although very effective in discrete manufacturing, ended up a fad in process plants because of poor application of the wrong statistical tools.

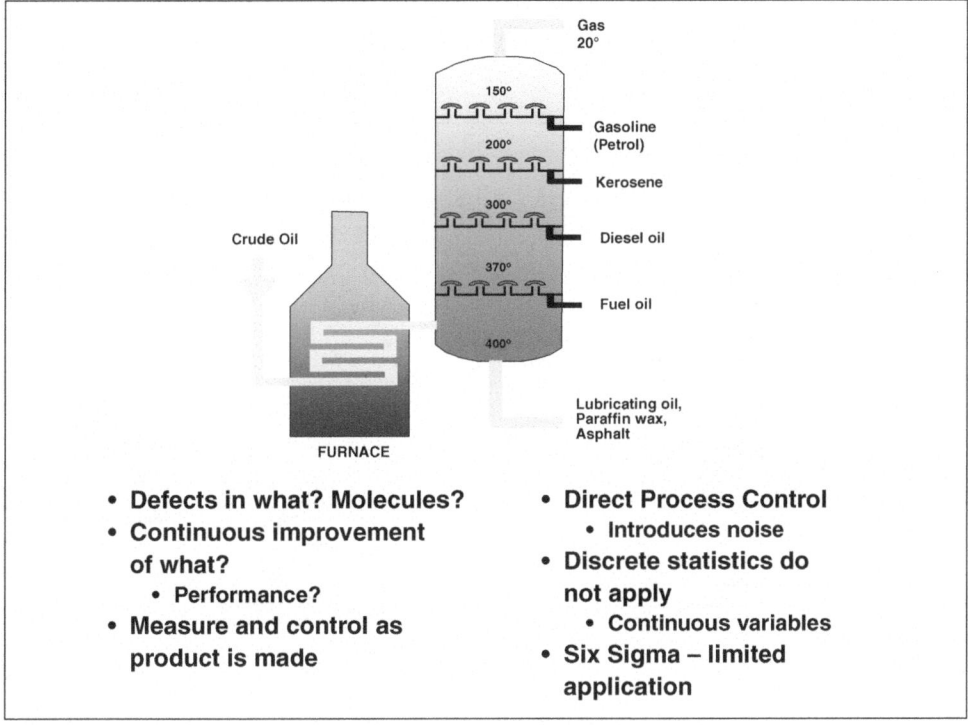

Figure 12-2 Quality Management in Process Manufacturing

The interesting aspect of this is statistical analysis is normally applied when direct mathematical conclusions cannot be determined. This is the situation in discrete manufacturing operations, but in process manufacturing operations, quality variables undergo direct or indirect measurement, and the advanced control methods previously discussed can deterministically control quality as well as production. In process manufacturing operations, reverting to statistics is often a step backward in quality control.

Today, most process manufacturers realize the direct process control they have been doing for decades can also do direct quality control. The key issue in making this happen is measuring the quality variables directly. As was pointed out, manufacturers could not initially measure a number of critical quality variables, so they had to send samples out for analysis. The problem was this introduced a significant delay in the control loop for the quality variables, since the operations and engineering professionals had to wait for the lab to complete the analysis. This could also delay any corrective actions. Since manufacturers were still making product as the lab analysis was underway, a quality problem could result in a considerable amount of off-spec product that could require rework. That ended up being expensive.

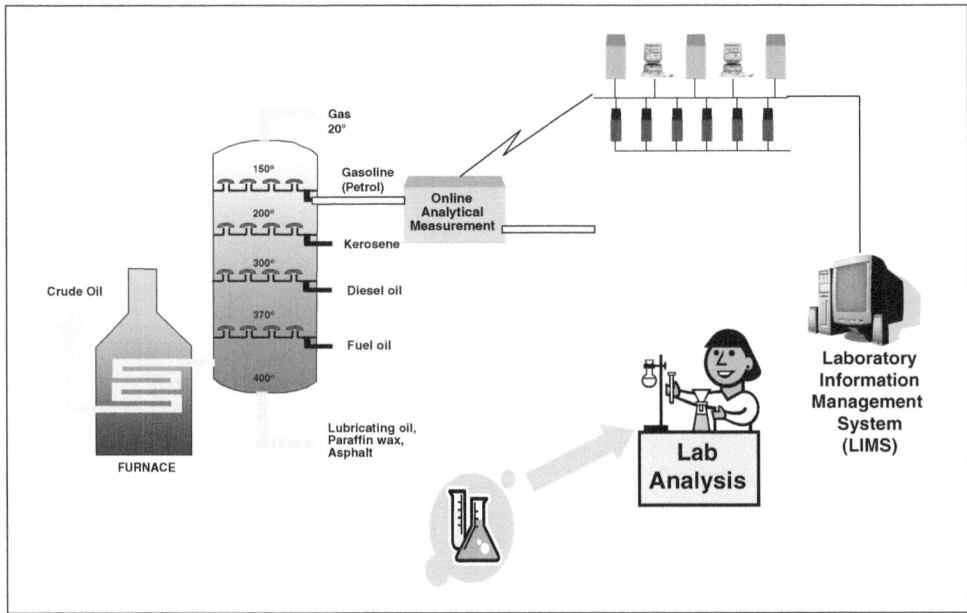

Figure 12-3 Process Quality Management

Over the past few decades, there has been a movement to reduce the dead time introduced by lab analysis into the quality control of process manufacturing operations. The first approach has been to implement computer-based lab analysis systems that could directly connect with DCSs (Figure 12-3). These computer-based laboratory systems are laboratory information management systems (LIMS) and they can significantly reduce the communication time between the lab and the operation. Reducing the dead time introduced by lab analysis makes the quality control problem much more manageable.

An additional and more effective way of attacking this problem has been to use online measurement approaches, such as analytical chemistry-based devices, to directly measure the quality variables that require control. Alternatively, software measurement techniques, such as factor analysis, process modeling, and mass and energy balances can infer the current measurement of quality variables with considerable success.

Quality Management is a very important function in any manufacturing operation. Producing product to spec is critical to the reputation of the manufacturer and to customer satisfaction with the products. Although considerable effort has gone into the development of quality management approaches and systems, there is still confusion with respect to the different approaches in discrete and process manufacturing. It is very important not to oversimplify the problem of managing quality by reducing it to the application of standard predefined tools, and it's equally important to make sure the problem is approached in a manner appropriate to the kind of manufacturing being done. Even to this day, statistical process control tools and approaches are being inappropriately applied to continuous process manufacturing operations by competent quality management professionals who learned their trade in discrete manufacturing environments. Hoped-for improvements will not occur unless the manufacturer applies the appropriate tools.

Review Questions

1. Briefly describe the differences in focus between quality management in discrete manufacturing operations and quality management in continuous process operations.

2. As a general rule, when should statistical process control be employed instead of deterministic process control?

3. What does the acronym LIMS stand for?

4. Why are online analytical measurement systems important to quality management in process plants?

5. Why does the concept of continuous defect reduction have limited applicability in process plants?

CHAPTER 13

Asset Management—Maintenance Management: Coming of Age

Sitting in a meeting with industry icon and president of Emerson Process Management John Berra, you can hear him say over and over again that unplanned downtime is a productivity and profit killer for a manufacturer. That is why a well-designed, well-executed asset management program is a solid hedge against any type of equipment breaking down at the wrong time.

"Asset management" is an interesting phrase in industrial operations because it assumes very different meanings depending on which people in the organization are using it. If a plant-level person is using the phrase, they are typically referring to the functions involved with maintaining the plant's capital assets (which include equipment, instrumentation, and automation and information systems). When the managers or executives of industrial companies use the phrase "asset management," they are most likely referring to maximizing the business value from all the capital and non-capital assets of the organization. For the purposes of this chapter, we will use the former meaning—the management of the maintenance of the capital assets in manufacturing and production operations. Therefore, we will use asset management and maintenance management interchangeably.

A Slow Turn

The science of asset management in industrial operations has progressed at a much slower rate than the science of control and process optimization. Up until the last couple of decades, the maintenance departments at industrial plants primarily repaired the equipment as efficiently and as quickly as possible on equipment breakdowns. This approach is reactive maintenance or "break-fix" maintenance. With reactive maintenance, one of the most important aspects of effective maintenance performance is having the right parts in stock so a technician can fix the breakdown quickly when it occurs. Therefore, in reactive maintenance environments, spare parts management became very important and since the cost of equipment downtime to the overall operation tended to be quite high, having a large stock of spare parts was the norm. Software systems for maintenance focused on the effective management of spares, including parts inventory management and purchasing. The software would also manage work order scheduling for the maintenance department. Quite often the measure of success of these systems came from minimizing spare parts inventory while still effectively responding to breakdowns.

Over time industrial organizations came to realize the negative impact of equipment breakdowns on the business performance of the plant was worth much more to the company than just the cost of the spare parts inventory. They also realized there had to be a better way. That is where preventive maintenance comes in. Preventive maintenance strategies calculate the expected operational time before a breakdown should occur in any piece of plant equipment, along with the impact of a breakdown.

Preventive maintenance also gave industry a couple of new acronyms. The maintenance team had to determine the average (or Mean) Time Between Failures (MTBF) of the equipment from historical records or vendor statistics and the Mean Time it would take To Repair (MTTR) the equipment upon failure. Statistics such as MTBF and MTTR could help predict failures and determine the consequences of the failures so a company could develop a more effective maintenance schedule.

Improving Uptime

With preventive maintenance, the idea is to schedule maintenance on the equipment within a time frame that will most likely enable the maintenance team to fix potential problems that might cause a breakdown, thereby preventing (or reducing the probability of) a breakdown. Doing this should improve the overall availability of the production equipment and greatly increase productivity. Maintenance software came together to develop schedules accordingly and to automatically create maintenance work orders.

In addition to these benefits, it often takes much less time to do preventive maintenance on equipment than it does to fix the problem after a breakdown has occurred. Take a natural gas processing operation, for example. Gas processing operations utilize large compressors that gather raw gas from the wells in the gas fields and feed it into the gas processing plant. At one gas plant, engineers found preventive maintenance on one of these compressors may take about three hours to complete, but if the compressor reaches the point at which it breaks down, the collateral damage caused by the breakdown could lead to the loss of the compressor for up to three days. The difference in production value from a compressor being lost for three days rather than three hours would amount to huge losses to the operation.

An additional benefit of preventive maintenance over reactive maintenance is the plant can schedule maintenance for a time that does not interfere with production operations. If a breakdown should occur while the production team is trying to complete a time-constrained order, they may not be able to meet their obligations, possibly resulting in performance penalties. If, on the other hand, preventive maintenance normally occurs on a scheduled basis, the maintenance team should be able to provide plant production management with a scheduling window during which the preventive maintenance needs to be done, and the production team can set up production schedules and short-term contracts accordingly. The net result is the business performance of the operation should significantly improve.

Although doing this is a good idea and should be a benefit to any plant that undertakes a preventive maintenance strategy, let's look at reality. Unfortunately, maintenance and production teams within plants often do not coordinate as they should and a schism forms between these teams that can be difficult to overcome. Some industrial operations have tried to address this schism by developing performance measurement systems with contextualized performance dashboards to encourage collaboration between operations and maintenance personnel. The results in terms of improved performance have been very promising.

With all its benefits, there is a downside to preventive maintenance. While there are calculations for the expected operational time of a piece of equipment before a breakdown, there are occurrences in which the maintenance team replaces components or equipment that really doesn't need replacing. That can carry its own costs in lost production and capital.

Maintenance by Probability

A parallel maintenance strategy that actually originated in government applications is reliability centered maintenance (RCM). This is a strategy for improving on the initial plant design in a manner that improves equipment reliability and reduces plant downtime. RCM can work in industrial operations, independent of whether they are using a reactive, preventive or some other strategy for the maintenance of the plant.

RCM professionals analyze the design, reliability, and failure impact of each major piece of equipment in an operation to determine the probability of a breakdown in the equipment over time and the business impact to the operation if a breakdown were to occur. If the impact is large, the most economical way to deal with it is to design an active backup of that piece of equipment into the plant. Upon failure, the backup can take over the job while they repair the primary. In this way, no production is lost to a breakdown.

Take a look at a large oil refinery that will shut down if a certain pump fails. Suppose the pump costs $300,000, but the lost production due to a single failure may be worth millions of dollars. The RCM engineer would evaluate the probability of a failure as well as the consequence to the business upon a failure and would most likely recommend installing a parallel pump that automatically switches on upon the breakdown of the primary pump. The cost of installing the backup pump is pretty significant, but the cost of not having one can be much greater.

One of the additional considerations the RCM professional would take into account before making a recommendation is the availability of the equipment for effective preventive maintenance. In the refinery example, since oil refineries operate continuously over a period of multiple years, shutting down an operating pump for preventive maintenance may not make business sense because shutting down the pump would mean the plant would have to be shut down as well. Not doing effective preventive maintenance on the pump would increase the risk of a pump breakdown. Parallel pumps may be the only reasonable way to mitigate this risk.

On the other hand, in a batch operation, a similar pump may move the finished product from a reactor to the storage tank at the end of a batch. If the batch cycle time for the products is fairly large, the discharge pump may only operate for 10% of the time the plant is operating. In this case, the RCM professional might not recommend the installation of a parallel pump since effective preventive maintenance can be done on the pump during the 90% of the time the pump is not required for processing. This might make the cost to install a backup pump unjustifiable.

Over the past decade or so, considerable work has gone into transforming maintenance systems to be more like maintenance control systems, in which instrumentation monitors key maintenance variables throughout the operation (a process called *condition monitoring*) and the maintenance system automatically responds. This is a predictive maintenance system. Predictive maintenance helps to close the maintenance loop in a fashion similar to the operation of feedback process control systems.

Predicting Failure

The condition monitoring instrumentation may be the same as the process instrumentation used to monitor and control the process, but there are additional, very specific measurements for maintenance, such as vibration, bearing temperature, oil temperature and the like that can provide a considerable amount of information that may predict an impending failure. A boatload of different companies have developed very sophisticated software to analyze this information so it can effectively predict impending failures. These predictions can often identify conditions that might lead to breakdowns weeks prior to the expected failure, giving maintenance and operations in the plant time to determine the best point at which to address the problem. These predictive maintenance approaches have significantly advanced the science of maintenance.

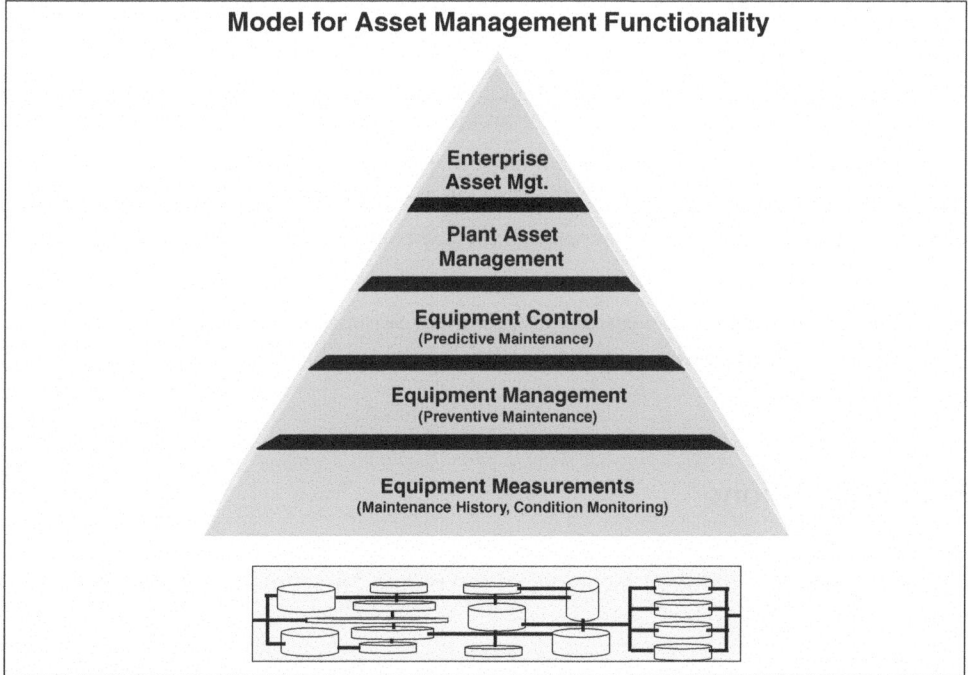

Figure 13-1 Functional Model for Asset Management

The model in Figure 13-1 provides a holistic perspective to the management of industrial assets, which has become increasingly sophisticated. This model shows the equipment measurement function as the lowest level function in the hierarchy. This function can either represent maintenance histories for each piece of equipment to establish preventive maintenance schedules, or direct condition monitoring for predictive maintenance strategies. The next higher level, equipment management, represents preventive maintenance strategies in which experience with the maintenance of the equipment can help the maintenance team develop reasonable preventive maintenance schedules to minimize the occurrence of failures.

The next level up, equipment control, uses the direct equipment measurements to predict impending failures in a predictive maintenance strategy. Above this is the plant asset management (PAM) layer, which represents the overall coordination of the various maintenance strategies used in industrial plants, including break-fix, preventive maintenance, predictive maintenance, and RCM, as well as maintenance scheduling and spare parts management. (It is not unusual or even unreasonable for different maintenance strategies to be employed for different equipment in the same plant.) Finally, the highest level on the model, enterprise asset management (EAM), represents those functions required to manage the assets across all the plants within an industrial enterprise.

The science behind industrial asset management has significantly grown over the last two decades and it continues to advance today. Considerable research is going into modeling the impact of effective collaboration between the operations and the maintenance of the same production assets in order to develop the optimal combined operations-maintenance strategy for each piece of equipment. These new collaborative approaches are starting to show significant progress in business performance improvement in industrial plants.

Review Questions

1. What are the two different meanings of the phrase "asset management" in industrial companies?

2. What do MTBF and MTTR refer to?

3. What does the phrase "reactive maintenance" refer to?

4. What does the phrase "preventive maintenance" refer to?

5. What does the phrase "predictive maintenance" refer to?

6. What is reliability centered maintenance?

7. What does the acronym PAM mean?

8. What does the acronym EAM mean?

CHAPTER 14

Human-Machine Interfacing— SCADA Software: Breaking Away from a DCS

Human-machine interfacing (HMI), when used in an industrial context, refers to the science of how personnel can effectively interact with industrial processes through automation technology. HMI is an inherent function in many automation systems, including distributed control systems (DCSs), but was not a major consideration for early programmable logic controllers (PLCs). HMI evolved into an independent category of industrial software in the late 1980s.

The original idea behind a DCS was for the system to be geographically and functionally distributable. Figure 14-1 presents some of the major functional components of a typical DCS. These functions include process control, logic control, field device interfacing to instruments and valves, PLC interfacing, advanced control, process information management (process historian and supporting software), human-machine interfacing, third-party intelligent device and system interfacing, advance applications, management information system (MIS) interfacing, batch management and process optimization, among others.

When a manufacturer acquired a DCS, they expected all of these and other functional components.

PLCs, on the other hand, were not as functionally rich. They were designed to be programmed from an engineering station and to run through their logic control sequences, reading the status of field instrumentation and driving switches through a field device interface. The operator interface could be fairly simple, perhaps only consisting of two functions (Figure 14-2). They often only had mechanical pushbutton switches as the operator interface and performed the same sequences over and over again.

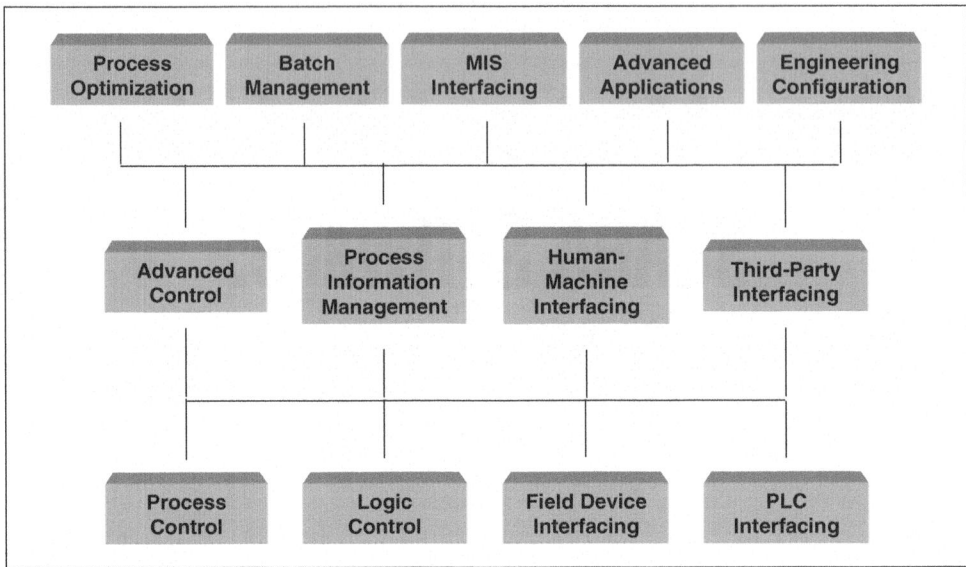

Figure 14-1 Functional Components of a DCS

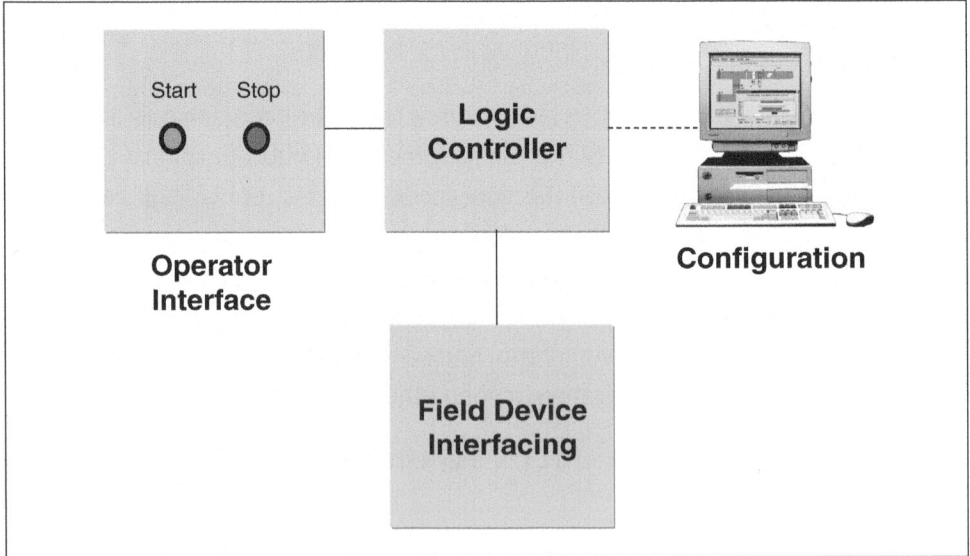

Figure 14-2 Functional Components of a PLC

As computers became smaller, less expensive and more capable, manufacturers using PLCs wanted to have computer-based consoles similar to those provided with DCSs to visualize the manufacturing process and to interact with it. Newly available personal computers provided the ideal platforms for such operator interfaces, and a number of software companies developed software to run in the PCs that provided a graphical operator interface for PLCs. The software designed to meet this need was man machine interface (MMI) or human-machine interface (HMI) software. The initial HMI software included a simple graphics engine with a graphics configurator and a simple historian software package for trending (Figure 14-3). Manufacturers saw value in putting this new HMI software on their PLCs and a new software market segment emerged.

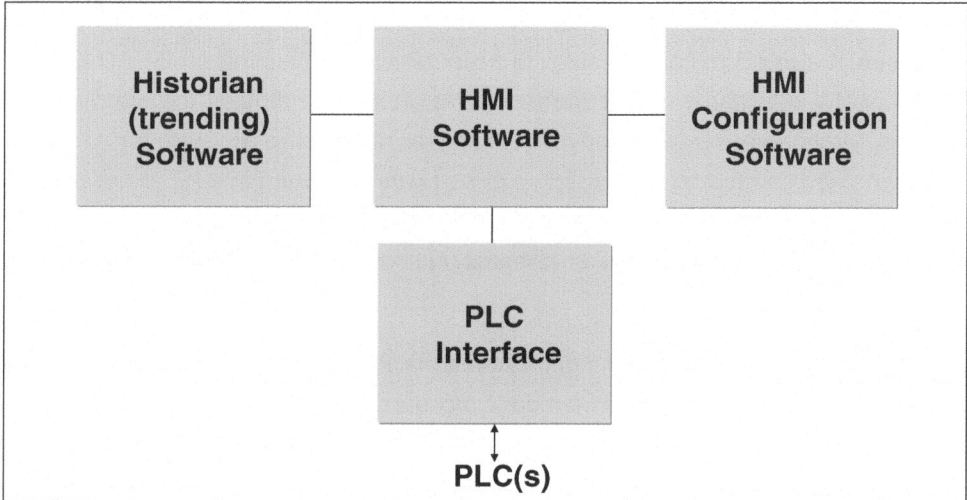

Figure 14-3 Functional Components of an HMI

The initial HMI software packages ran on PCs using MS-DOS as the operating system. Although there were a number of well-received packages, the weaknesses of the operating system stifled the general acceptance of HMI software. With the introduction of Windows by Microsoft, the HMI software market took off, led by software companies such as Wonderware, Intellution, Iconex, Intech Controls, and Citect.

Some of the HMI software companies began to provide additional functions formerly found only in DCSs, primarily process control and the associated engineering configuration software. With this increased functionality going well beyond basic HMI, the software running on these systems started to be referred to as supervisory control and data acquisition (SCADA) software. The process control component of these SCADA software packages operated in the PC and connected to the process variables through the PLCs using their analog input and output capability. Note, however, although the control software in such SCADA systems was capable, manufacturers were slow to use it for direct control of their processes due to the unreliability of PCs and the lack of real-time capability in the Windows operating system.

PC software providers encouraged calling it SCADA software because it distinguished the software from simple HMI systems and allowed them to position the control software in their packages as supervisory rather than direct process control. That move improved its acceptability in the marketplace. In addition, this HMI-SCADA software was designed to be simple to set up and use. This ease of use resulted in a rapid increase in the popularity of HMI-SCADA software. As we have seen, it also led to considerable confusion as to the meaning of "SCADA." As SCADA software became more common it was often quite difficult to determine whether someone using the acronym SCADA was referring to a system for measuring and controlling over distance or software operating in a PC. It became even more confusing as PCs operating SCADA software developed into effective solutions for SCADA system master terminal units or MTUs (covered in Chapter 11 on SCADA systems).

This software was initially on tap for traditional PLC-based manufacturing operations, but as PLC vendors added basic process control functions to the software, it was not unusual for PC-based systems with SCADA software to run in applications traditionally reserved for DCSs. DCSs offered advanced process control and optimization functions enabling much more sophisticated control of continuous processes, but manufacturers found that there were a number of processes that might not require the sophistication of a DCS, so SCADA software with PLCs served the purpose.

Over a very short period of time, the capability of PCs increased dramatically while the price decreased. As this happened, SCADA software companies added more and more functionality to their software suites as shown in Figure 14-4, and the line between DCS applications and PLC applications started to blur. Batch management software, advanced applications, interfaces to all kinds of intelligent systems and devices, statistical control packages, and more advanced historian and information management functions all found their way into SCADA software. The basic functionality of a PC-based SCADA system became very similar to that of a DCS. In parallel with this functional growth in SCADA software, DCS companies discovered the power of personal computers and started implementing traditional DCS software in PCs, making PCs integral components of DCSs. Today it is somewhat difficult to distinguish between some of the more advanced SCADA software systems and some of the more progressive DCSs.

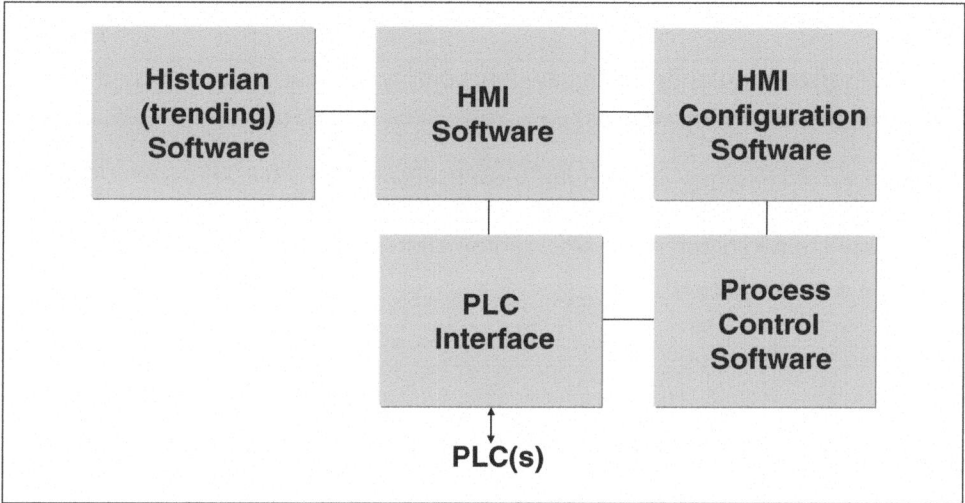

Figure 14-4 Functional Components of SCADA Software

Today, DCSs still have the advantage when it comes to process control and advanced process control; they also tend to hit higher levels of reliability than SCADA software. SCADA software, on the other hand, often holds the advantage in terms of ease of use. This is probably because this software has not expanded to take on some of the more complicated and scientifically difficult problems DCSs have, but also because the designers of SCADA software have traditionally concentrated on making their software simple.

Digital computer and software technologies are advancing to the point at which systems initially designed for very different, but complementary, industrial automation requirements are converging. The lines between these two classes of industrial automation systems will continue to blur and should eventually disappear as more universal automation platforms emerge.

Review Questions

1. List at least four of the major functional components commonly part of a DCS.

2. What intelligent devices were HMIs initially designed to work with?

3. As HMI software expanded with new functionality, what acronym was used to describe these software systems?

4. What type of computer enabled the cost-effective use of HMI software?

5. What are the two different classes of automation technology that the acronym SCADA is associated with?

CHAPTER 15

Manufacturing Execution Systems: Two Becoming One

As computer-based industrial automation systems were evolving, computer technologies also expanded within the business management of industrial and other corporations. It soon seemed as if there were two separate worlds within one industrial company: The manufacturing unit and everyone else.

Business systems were typically the domain of the Information Technology (IT) organization which was led by a Chief Information Officer (CIO) reporting at the executive levels. The position of the CIO pointed to the importance of these systems to the executive management of industrial operations. On the other hand, the plant engineering, maintenance, and operations teams managed the plant-level industrial automation systems.

Over time, as computer systems evolved, the technologies used to implement business systems and those on the industrial automation side appeared to converge. It became difficult to clearly distinguish between the technologies used in each. People in manufacturing operations believed value might be added to their businesses by connecting the business systems and the industrial systems. Since these two classes of systems worked on common technologies, bringing the business world and the industrial world together seemed to be a reasonable objective. Reasonable yes, but as they say, easier said than done.

Starting in the late 1970s, the first significant attempt to bring the two worlds together through technology came under the banner of computer integrated manufacturing (CIM). Most CIM implementations were technological approaches to business problems. Minicomputers had been introduced and were being used for business and industrial automation, and appeared to provide the perfect intermediate technology between the two systems. Most DCS companies developed computer gateways to connect into the most popular minicomputers of the day (Figure 15-1).

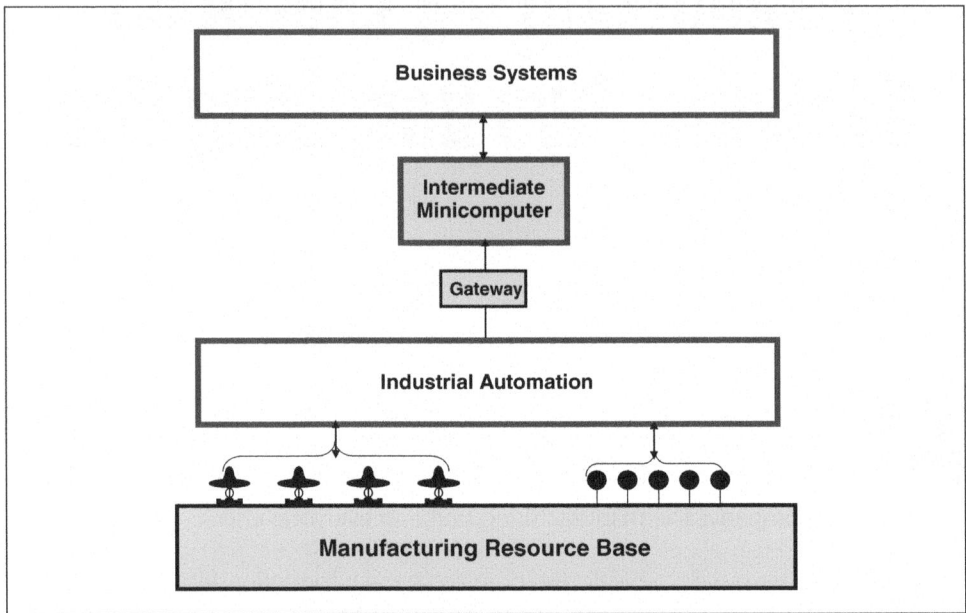

Figure 15-1 Computer Integrated Manufacturing

The gateways passed as much process data as possible in a given time period, up to the intermediate minicomputer. The data was on a circular (i.e., revolving) file of a specific size on the intermediate minicomputer, which would continually overwrite older information in the file as new information came in. This essentially provided a time-based snapshot of real-time plant data to the business systems. From a DCS perspective, this approach appeared to be reasonable because the plant data was made available to the business systems. But the business systems hardly ever accessed the data because the IT organization seldom understood either the content or the context of the data. CIM implementations seldom realized the desired results.

Although the technical approach to CIM was almost universally a failure, a number of innovators saw an opportunity. DCSs provided a considerable amount of standard functionality for plant-floor automation, but there were a number of functions not offered by DCS suppliers that manufacturers wanted. And if a customer wants something, someone will deliver. As a result, a bunch of innovative software companies started to develop applications designed to run on standard minicomputers, such as those offered by Digital Equipment Corporation and Hewlett-Packard. Since computer gateways had already been developed for connecting most standard minicomputers to most major DCSs, these new software applications were viewed as extensions of the DCSs.

As more and more of these applications hit industry, a new technology domain started to emerge between industrial automation and the business side. Advanced Manufacturing Research coined the name manufacturing execution systems (MESs) for this domain between automation and business systems, apparently because there were so many different types and flavors of applications that might fit into this space that it was difficult to categorize (Figure 15-2).

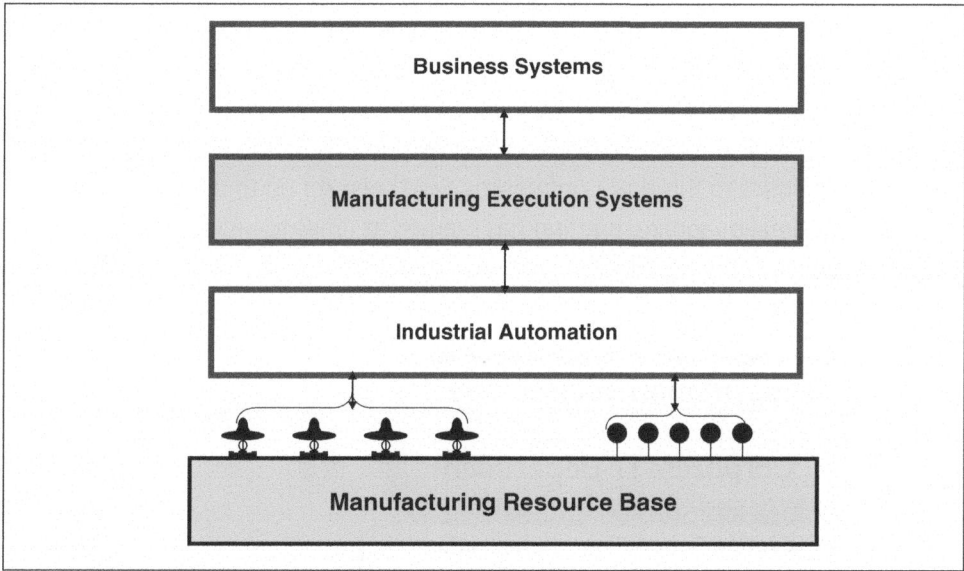

Figure 15-2 Manufacturing Execution Systems

As the applications in this MES domain started to gain acceptance, a three-way battle for the MES space started. To try to gain sales the MES companies were garnering, the business system companies and the DCS companies started adding similar functionality to their systems.

This resulted in a very interesting internal organizational battle in industrial companies. As we have seen, the DCSs were typically managed by the plant engineering departments, while the business systems were managed by corporate IT. Although both of these groups used computers as their primary technology platform, they performed different functions within their organizations and had been able to operate independently from each other up to this point. With the battle for the MES space, the IT departments and plant engineering departments often found that they acquired software to enable their companies to do essentially the same thing.

These two technology-based organizations were suddenly forced to confront each other. The technology focus of these two groups was so different that they often did not understand what each other was saying and found themselves in conflict with each other. The conflict grew into distrust and fear. While the two sides had similar goals, their languages were completely different; therefore, communications were poor.

In addition, the plant automation teams knew that the CIO who headed up the IT group was on the corporate executive board and had organizational power, and they feared being absorbed into IT. The IT teams knew the plant teams worked with computers but had no knowledge of real-time systems, or process control, or the advanced engineering of the plant engineering teams and tended to avoid the plant-floor systems. A schism formed between these two groups that in many industrial companies continues to this day. Conferences have been set up to discuss solutions to the departmental quagmire, but that has seemed at times to be very similar to negotiating peace in the Middle East.

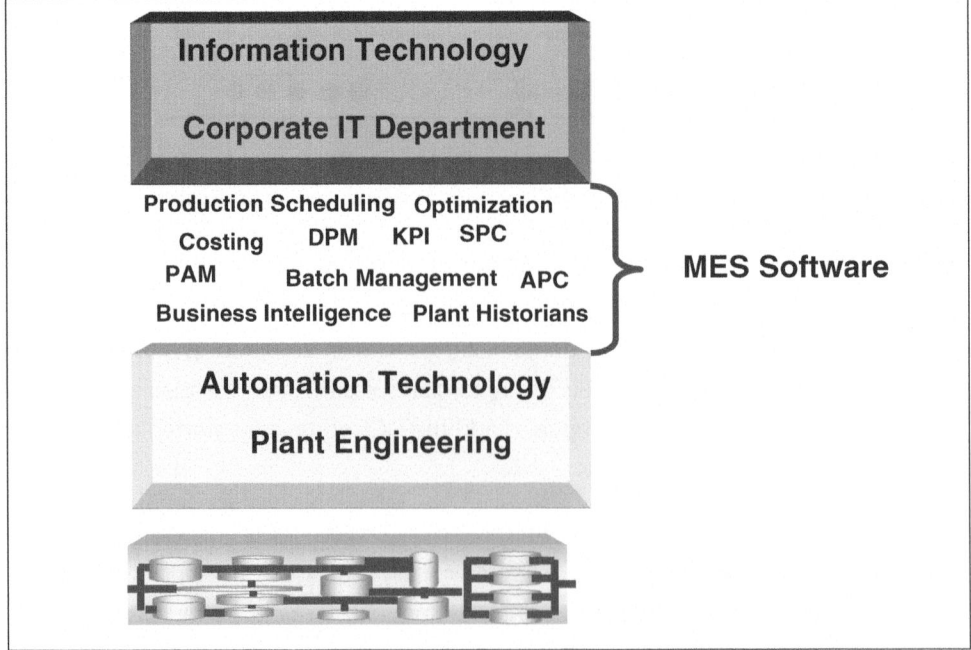

Figure 15-3 The Manufacturing Execution System Software

The net result is the MES domain is really a set of software applications developed by small software companies. Some very smart companies developed some successful applications at the MES level that have significantly changed the landscape of industrial computing (Figure 15-3). OSI Soft introduced a process information management systems (PIMS) called PI that provided a number of information management tools around a process historian and could work with most DCSs. This allowed manufacturers to build a common process information system even if there were automation systems from multiple vendors. Also, human-machine interface (HMI) companies (e.g., Wonderware) acquired a number of MES software companies and built suites of products that significantly extended the functionality of their systems. Needless to say, the MES domain was, and is, very dynamic.

There is an ongoing battle over the MES space among the industrial automation and business system companies, as well as between the plant engineering and IT organizations serving industrial companies (Figure 15-4). Industry consultants have claimed that the MES space has effectively disappeared as the automation systems and business systems have expanded their functional space to cover traditional MES functions. But consultants always say that kind of thing. The reality is that as long as there are unmet needs in industrial operations, entrepreneurs will develop new applications to meet the needs and will most likely develop them to be independent from the automation systems and the business systems. This is how MES got started, and it should continue that way.

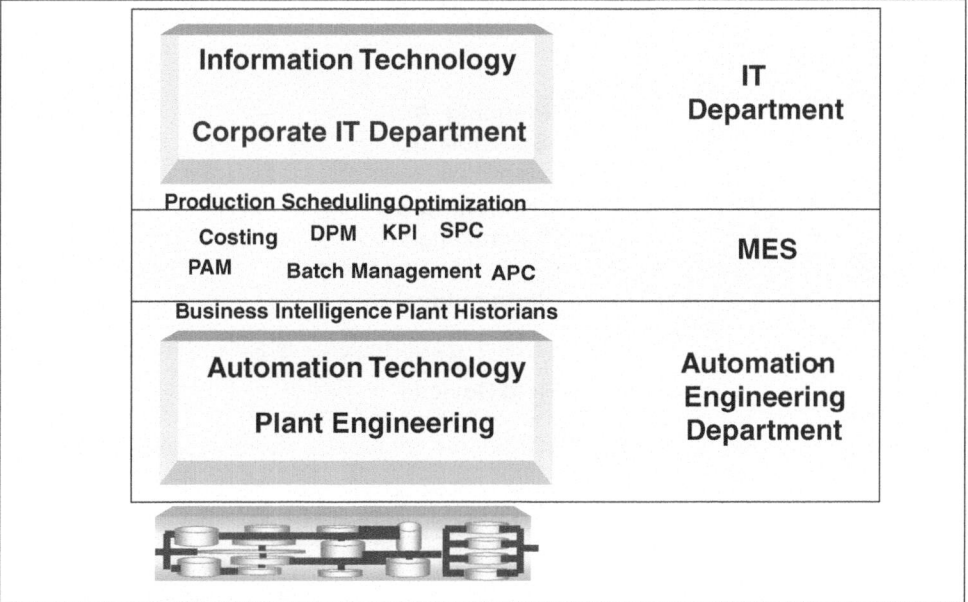

Figure 15-4 The Manufacturing Execution System Space

The MES domain has served two very important functions in industrial automation: It has served to rapidly expand functionality in a market space that had traditionally moved at a snail's pace, and it has served to start the process of having the two technology organizations within industrial companies, IT and engineering, learn how to deal with each other.

The initial dealings between these two departments have been difficult, but as the initial fear has started to subside, true progress has started to take place in joining these two organizations in a manner that actually helps drive business value. There are some industrial companies that have decided to combine their IT and engineering teams and believe it or not, the world did not end. The results were very positive.

As business systems and industrial automation systems started to come together, there was considerable confusion over what functions would best be performed in the various technical architectures. During the 1980s, a team of industry specialists headed by Dr. Theodore Williams of Purdue University worked on defining a technology model of the functions from the plant floor up through the enterprise business systems. The result was an extensive functional model, developed using functional analysis tools, which essentially partitioned the functions into five horizontal layers and detailed the specific functions at each layer. This model is the Purdue Reference Model, and people throughout industry still use it today (Figure 15-5).

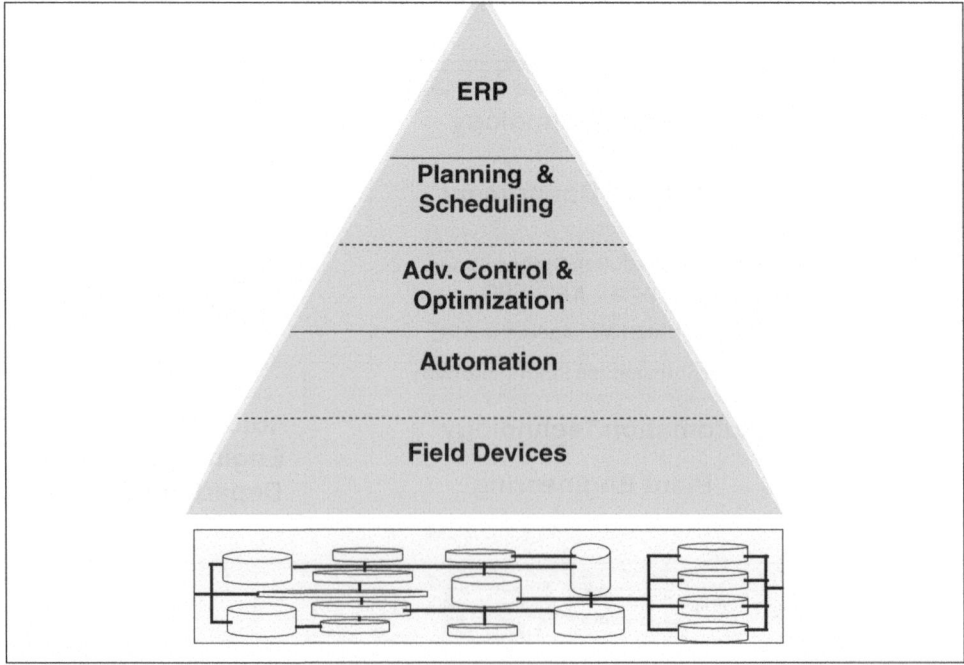

Figure 15-5 Purdue Reference Model

This overview of the Purdue Reference Model can be very deceptive in its simplicity. The important point to realize is each of the five levels of functionality shown is backed up by very rigorous and detailed functional models. The depth of this work should serve to demonstrate the level of complexity involved in integrating industrial automation from the plant floor to the enterprise.

Review Questions

1. What two classes of systems does MES software normally operate between?

2. What does MES stand for?

3. Why was this software called MES?

4. What class of computer system initially made MES software viable?

5. Name five functions traditionally found in MES software.

6. What university provided leadership in the development of the CIM reference model?

CHAPTER 16

Enterprise Resource Planning: Business Software on Top of Automation

Born on the business side of the tracks, enterprise resource planning (ERP) has had a huge impact on plant automation. As today's business systems and automation systems start to converge, professionals on both sides need to understand the technological and organizational implications of each others' domains.

When the computer was first introduced "way back when" as a potential tool to help manage the information of business operations, only the largest organizations could afford to acquire and program a computer. Computers were extremely expensive, not easy to program, and not easy to operate. Over time the cost of computers declined, and new tools made the programming and operation of computers much easier. U.S. Navy Admiral Grace Hopper developed perhaps the most important of these tools, the A-O compiler. It, and its successors, enabled the programming of a computer with high-level, human readable computer languages that could compile down to an executable program that could operate within the computer. That work resulted in a number of application specific high-level languages, such as the Formula Translation Language (FORTRAN), for scientific applications and the Common Business Oriented Language (COBOL) for business applications. This innovation made it easier to train effective programmers.

The combination of these tools and declining prices made it feasible for a number of businesses to acquire computers and to hire people to program the computers to perform some of the major functions of the organization. As the value of the computer became apparent, more and more programs were developed, and each of the major functions of the business, such as payroll, accounts receivable, accounts payable, human resource management, finance, accounting, and sales typically had multiple programs developed and stored in a library, to run whenever they were needed. The net result was there were often hundreds of different programs stored in any company's library. This presented huge organizational challenges.

To make matters even more challenging, the field of computer programming was in its infancy and the storage capacity of the early computers was very constrained. Programmers developed the programs so they would fit in the computer's memory in order to run. This often required the use of programming techniques that might be difficult for another programmer to understand. There was no such thing as standardized programming, and it was typically very difficult for one programmer to modify a program developed by another programmer. If a programmer who created a parcel of programs left the company, there was a problem.

The net result was as the business grew and expanded, the computer programs developed by long departed programmers did not expand along with the business. This often led to frustration within the programming departments, who were treading water, modifying programs developed years earlier, and trying to keep up with the needs of the folks who wanted better and more timely information to run the business in an appropriate manner (Figure 16-1). The computer programs designed to help the businesses run better had become an anchor to business operations, preventing them from moving forward. The cost to maintain and modify business programs started to skyrocket. Programming staffs swelled. Internal Information Technology (IT) departments suffered devaluation by business executives. An information management crisis had developed. Companies needed a standard way to do business.

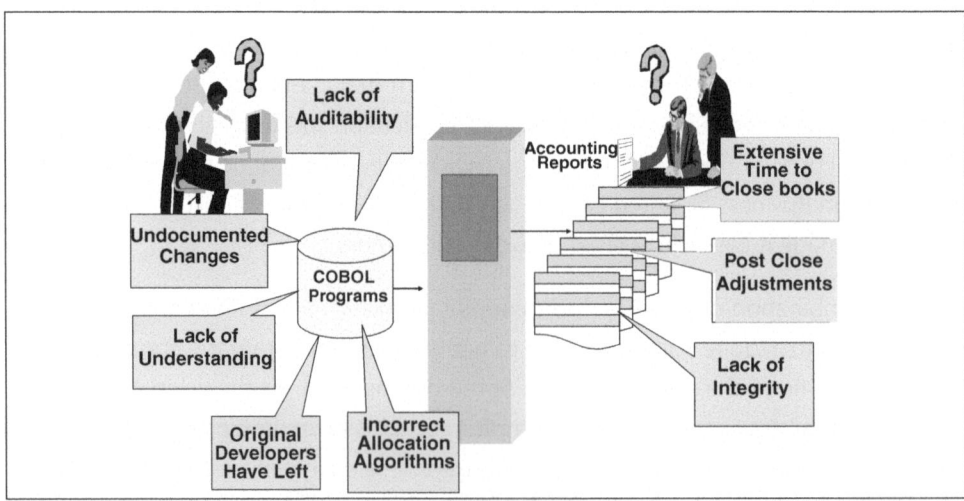

Figure 16-1 Early Computer-Based Business Systems

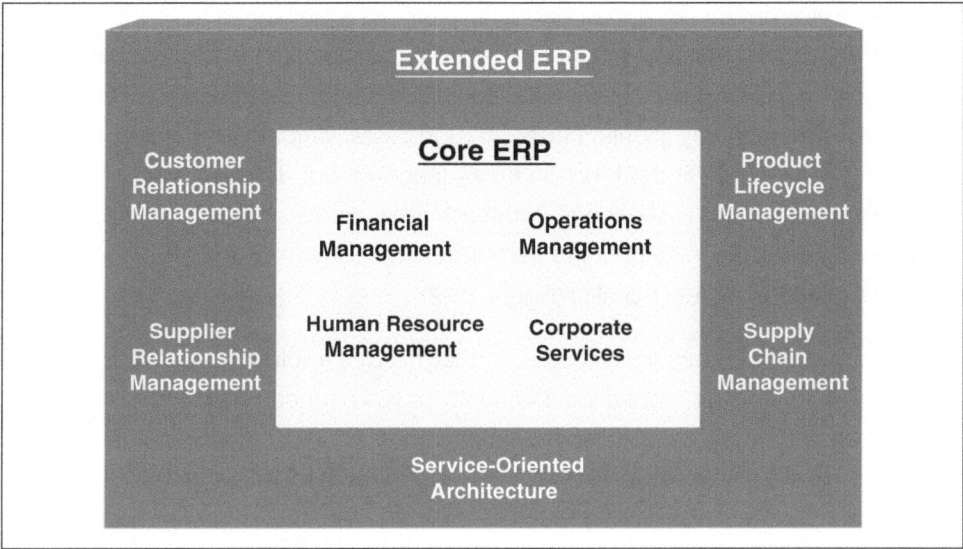

Figure 16-2 Enterprise Resource Planning

Clearly there was a huge need in the marketplace, and a few entrepreneurial software companies hit the market with new software. The initial approach of most of these companies was to develop programs or systems of programs to cover the major functions most businesses had to perform as part of their normal operations. Since the number of functions normally performed in any company's business information management system is quite large, the software companies selected a critical subset of the overall functionality, such as financial management, human resource management, operations management, or other key corporate services as a starting point. The unique selling proposition for this new class of software was not based on new capability; rather it focused on providing standard and maintainable software systems that performed essentially the same functions as the traditional home grown COBOL programs, but did not require the huge amounts of overhead to support them. This was the beginning of ERP software, and companies such as SAP, Oracle, and Baan were among the many that led the way.

Since the demand for maintainable business software to replace non-maintainable home grown software was very strong, the ERP companies became successful in a short period of time. The clamor was unprecedented. As the success of the initial ERP providers became evident, entrepreneurial software companies started to develop products to cover business functionality not included in the initial ERP software suites, such as sales management, customer relationship management, supply chain management and the like. Some of the early successful ERP suppliers found themselves in the position of having to either develop software for these functions or acquire companies that already had proven product in these areas. Acquisitions led to ERP portfolios partially developed within the software companies and partially developed by acquired firms. In most cases, various components of these portfolios were effective at the function they supported, but different functions across the portfolio tended not to work well with each other. Some of the ERP companies marketed to this situation by defining the various components of their functionality as core and extended ERP functionality (Figure 16-2).

A few leading ERP suppliers recognized they needed a solution that would allow software components developed independently to work together as though designed as a single system from inception. Standard computer networks that permitted different computers to interoperate had been important in enabling new levels of integration, but they were not enough to meet the application interoperability requirements of the business software systems. ERP suppliers needed a new approach that went beyond simple networking to offer systems and application services, such as naming services, visualization services and database services separate from the applications but available to the applications in standard formats.

That technology is service-oriented architecture (SOA). SAP's NetWeaver is a good example of a business SOA that enabled all the software developed within SAP and acquired by SAP to operate as a single system. Some SOAs today use visualization services and other services developed for web-based operations, which can make them easier for third-party applications to utilize in certain situations. The introduction of SOAs, primarily by the ERP suppliers, has certainly led to major advances in application interoperation.

One of the more interesting impacts of the availability of ERP software was dramatic changes in IT organizations. IT organizations had traditionally consisted of large numbers of computer programmers under the guidance of a few systems analysts. The analysts would design the software systems and the programmers would implement them. With the advent of more standard ERP software, the requirement for computer programmers declined. The IT organizations had to hire or develop talent that understood how the ERP software worked so they could configure it for use in their companies. Many businesses hired consultants to implement their ERP software. This relegated the IT organization to a predominantly maintenance and adjustment function.

Largely as a result of this, there was, and still is, an ongoing debate within many businesses about whether it is more appropriate to maintain an internal IT organization or to outsource the function altogether. Some businesses have started to recognize this decision may not be a simple one, and there are some aspects of IT that may be unique and strategic to the business, while other aspects may be more like commodities. As a result, some businesses have partitioned their IT organizations into two groups: the specialty functions, under a heading such as Information Management, and the commodity functions, typically under a traditional IT heading. This allowed a new level of discussion often focused on outsourcing IT, but maintaining an internal Information Management organization.

Over time, the advent of ERP systems and the changes within them have had a significant impact on plant automation. As business systems and automation systems are starting to converge, it is important that IT/Information Management and plant automation specialists understand the technology and organizational implications of each others' domains. This understanding will lead to better communication, which has become even more critical for industrial businesses as they are pulling the IT/Information Management and automation engineering teams into the same organization to try to generate commonality and synergy between these two technology-based functions. As this trend continues, new approaches and solutions will evolve to help industrial businesses drive to even higher levels of value.

Review Questions

1. What does ERP stand for?

2. Name two ERP companies.

3. Prior to ERP software being available, business programs were developed by each industrial company primarily in what computer programming language?

4. What are four core functions of most ERP software?

5. What are two extended functions often provided with ERP software?

6. What does SOA stand for and what does it do?

CHAPTER 17

Automation System Integration: Enabling the Right Solution

Integration in and of itself adds no value; solving real business problems does.

The need to get automation components and systems from different suppliers to connect together became apparent as soon as the digital computer became the basis for automation system design.

Although the programmable logic controller (PLC) was originally a solution for the control of discrete manufacturing operations, manufacturers also found PLCs useful for the control of operations in process plants. Process computers and PLCs soon saw use in plants, each controlling separate, but related sections of the operation. When manufacturers found they needed coordination of control between the plant sections, it drove the need for communication between different process computers and between process computers and PLCs, as well as for communication with other intelligent devices.

In the case of PLC to process computer communication, if all that was required was the connection of a few discrete field inputs and outputs, the first course of action was often to connect the outputs from the PLC input/output (I/O) module to inputs on the process computer I/O module and vice versa (Figure 17-1). This type of connectivity offered an approach plant engineers or electricians familiar with wiring control systems and real-time data transfer could easily implement. The downside to this approach was a connection had to be set up for each field value being communicated between the systems. That approach could become quite costly as the number of values increases, and the only information communicated between systems is I/O data from field devices.

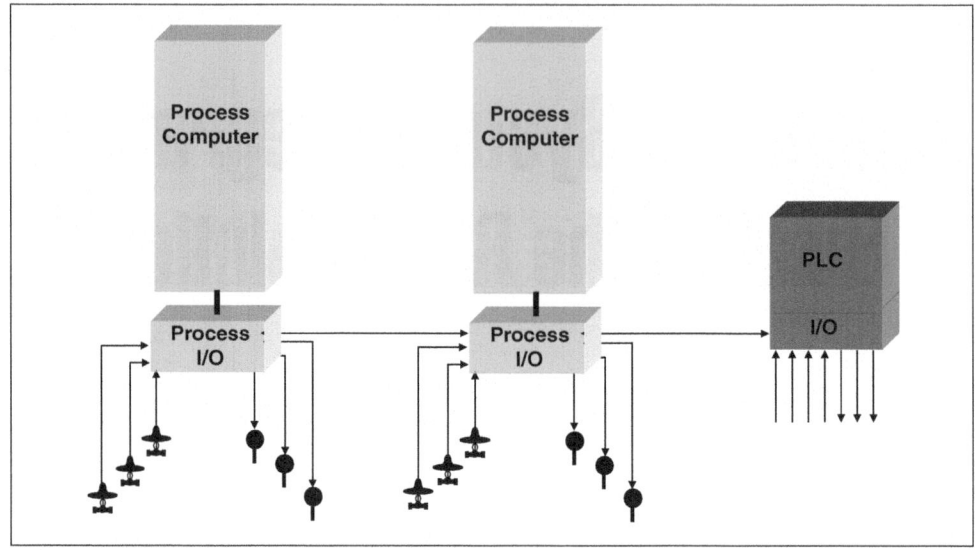

Figure 17-1 Automation Connectivity through I/O

As the volume of data to transfer between systems increased and as more data types other than just field inputs and outputs needed to be transmitted, manufacturers sought a more robust approach to communication between systems. Most computers had available built-in communication ports initially designed for communicating to peripheral devices, such as printers or terminals, that could handle different data types and fairly high volumes of data flow. Communicating between systems by directly connecting a communication port of one system to that of a second system became the obvious solution (Figure 17-2). Unfortunately, at this phase in the development of software for computers, the programs to direct the dataflow between systems typically had to undergo custom development for each application and for each system. Developing custom communication software required specialized knowledge and skills, which led to the creation of a new class of service providers called automation systems integrators.

As process computers evolved into distributed control systems (DCS), more sophisticated and intentional approaches to automation systems integration came into play. DCS suppliers designed specific functional modules as gateways to enable integration with other automation systems, such as PLCs, and intelligent devices (Figure 17-3). These modules included (somewhat) preprogrammed basic communication software in order to reduce the customization requirements and to aid in the setup of these interfaces for communication with specific devices and systems. Although these modules were easier to configure than the communication ports had been, they were quite expensive and still required an automation systems integrator to configure them and program them correctly. At this point, many DCS suppliers developed their own systems integration services, focused primarily on integrating third-party intelligent systems and devices into their own systems and essentially going into competition with independent systems integrators.

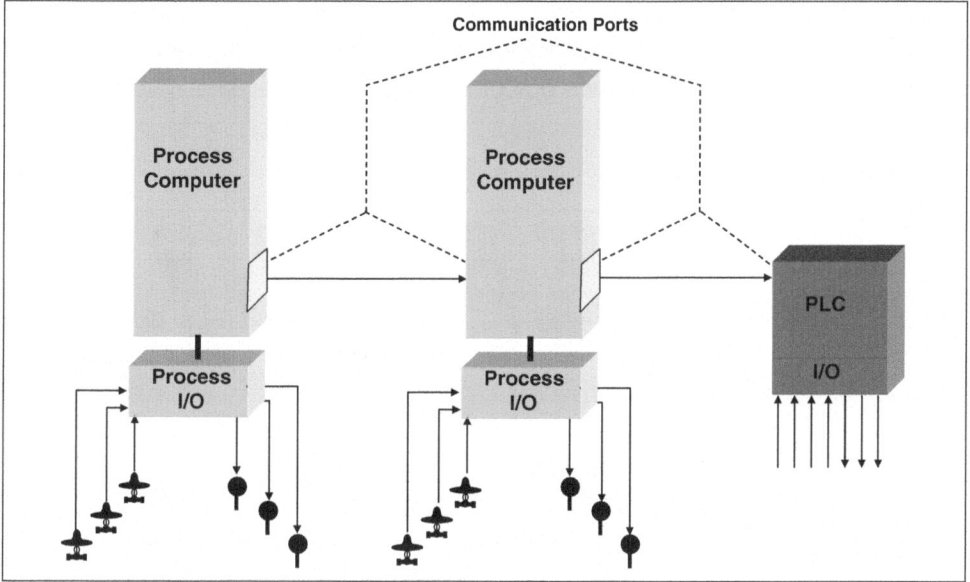

Figure 17-2 Automation System Connectivity through Communication Ports

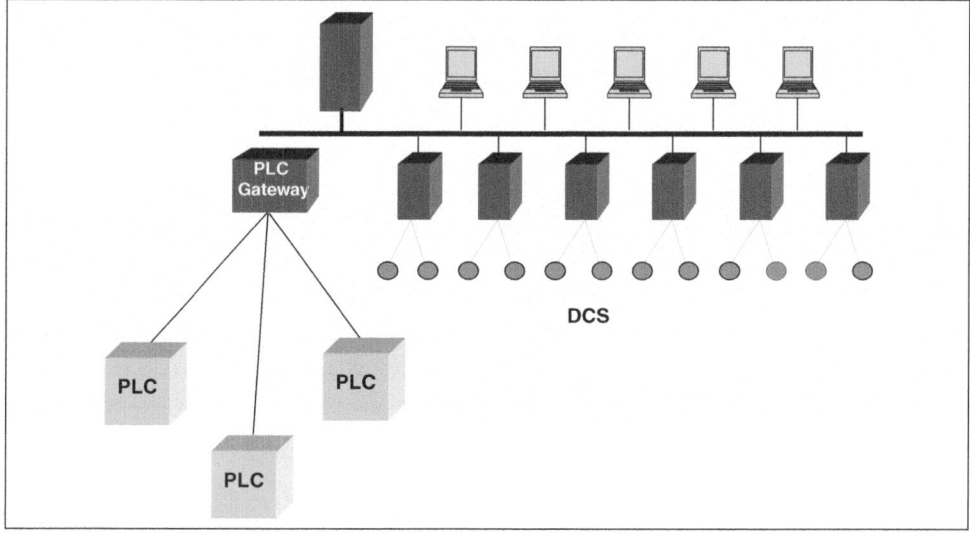

Figure 17-3 DCS Communication Modules

The next phase in automation systems integration, the integration of automation and business systems, began when industrial companies realized there were huge amounts of data generated from the plant-floor operation contained in the plant automation systems, which could be of value in running the business more effectively. To get this data from the DCS to the business systems, DCS companies developed computer gateway modules (Figure 17-4) designed to connect into a communication port of one of the more common minicomputers of the day. Since these minicomputers connected into business networks as part of the business computing infrastructure, providing process data to these minicomputers made operations data available to the business computing environment.

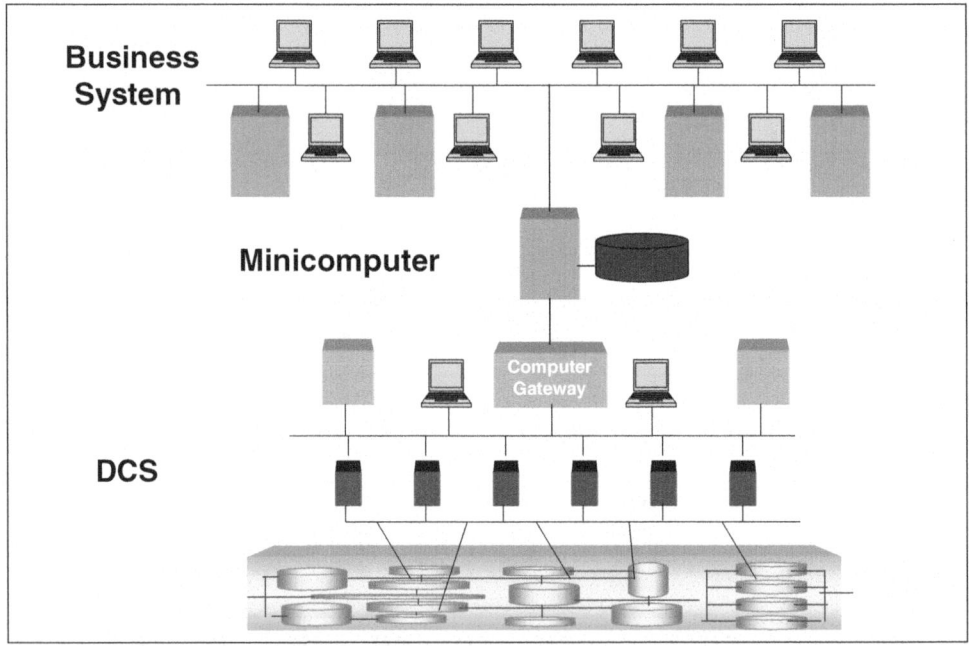

Figure 17-4 DCS Computer Gateway

DCS suppliers typically developed data storage software for the minicomputer in order to be able to store the plant data in an effective manner. This software stored as much process data as possible on the limited bulk storage available with the minicomputer. To accomplish this, the software in the minicomputer would request transfer of the collection of process data once every minute, or over some other time period, and would store the collected data set as a record in a file in the storage system. The process data might either be collected values from field instruments or averages for the time period. During the next time period, the system would collect another data set and that would go in the next record of the file. This data transfer would continue until the storage resources of the minicomputer were full and then the process would start over with the first record of the file being written over the data the computer had previously stored in that record. As has been mentioned in earlier chapters, this type of data collection approach is called the circular file approach, because the records are overwritten in a somewhat circular manner.

With this type of data collection system, the minicomputer's bulk storage would contain a snapshot of process data for the period of time the size of the storage device would allow. It was not unusual to see systems set up like this with thousands of process data values in each record and with enough records to cover the last week. Since the communications were typically only from the DCS to the minicomputer, and large amounts of process data were sucked across the computer gateway and into the minicomputer, this approach was known as the vacuum cleaner approach (Figure 17-5). The minicomputer was the vacuum cleaner sucking all the data it could out of the DCS. Once again, specialty talent implemented these systems, and systems integrators and the company's DCS supplier began offering such services.

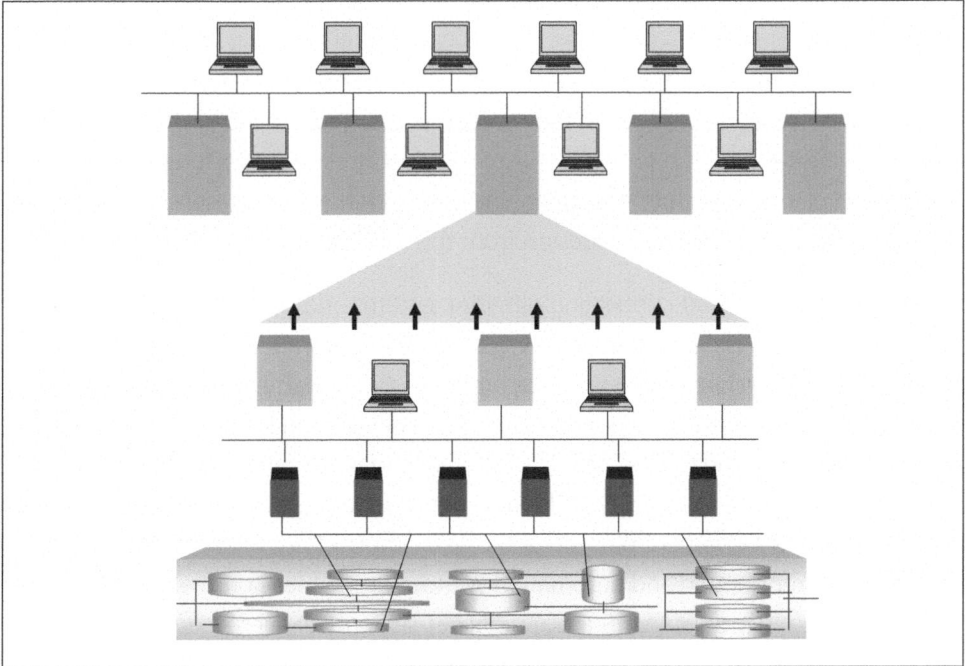

Figure 17-5 The Vacuum Cleaner Approach

Unfortunately, a survey conducted around the time of these systems found most of the data that had been sucked up into the minicomputer was not even touched by the business system before it was overwritten. The reason was to the business systems professionals, the data in the minicomputer was nothing more than useless raw data. The automation professionals provided as much raw data as they could without much context associated with the data, and the business systems professionals never really understood the data nor did they know what to do with it. This initial attempt to integrate the business and automation domains is important because it points out the naïveté of the automation teams and the information technology teams about what actually took place in the other's domain. This naïveté is still very much in place today.

Following quickly on the heels of the DCS gateway approach was a broader and more sophisticated-sounding movement to encourage the convergence of all computer-based domains under the label of Computer Integrated Manufacturing (CIM). CIM was a movement that virtually sprang up overnight. As soon as it hit the market, it seemed every industrial player wanted to have CIM. Unfortunately, CIM was more slogan than substance. Business managers typically viewed CIM as providing the information they required in order to run the business, while technologists viewed CIM as a way of meeting managements requests for data and getting on with other important work. It is not that people did not know what CIM was; everybody seemed to have a strong definition. It was just that no two definitions matched. Industrial companies spent millions of dollars connecting every intelligent system and device in their operation together, and when they had accomplished this feat, nothing seemed to work any better than it had before. Systems integrators made fortunes during the CIM era and plants were not performing any better. Figure 17-6 presents an overview of an actual CIM implementation design from this era.

The problem was that connecting different systems and devices together does not solve any business problems. Rather, it only overcomes communication barriers initially introduced by the technology itself. Overcoming technical barriers is an important first step, but business value improvement comes from developing solutions that address business issues once the barriers fall. The latter step was seldom made. Integration is valueless in and of itself, and only adds value if it is done to solve some larger business problem. The CIM movement presents a classic example of an abnormal focus on the technology of information and automation without much focus on the business of manufacturing and production. Such approaches seldom—if ever—add value to the business. So much money was spent on integration during this era with no visible benefits that many industrial business managers became jaded about spending on automation and business information technology.

Somewhat in parallel with the CIM movement, there was a corresponding movement in the development of communication standards. The idea was standard ways of communicating between systems would significantly reduce the cost and effort involved in integrating systems together. One of the first efforts in this regard was the Manufacturing Automation Protocol (MAP) sponsored by General Motors. GM was one of the world's largest purchasers of automation technology and had considerable clout in the industry as a result. They saw value in connecting their automation and business systems together and attempted to drive the MAP standard throughout the automation industry. MAP never really caught on across the larger manufacturing industry and this movement slowly died out. The communication standard that did catch on across industry during the 1980s was the predefined Ethernet standard. Ethernet had many of the characteristics required in industrial environments and had proven stability, and is the basis of most computer-to-computer industrial and business networks in use today.

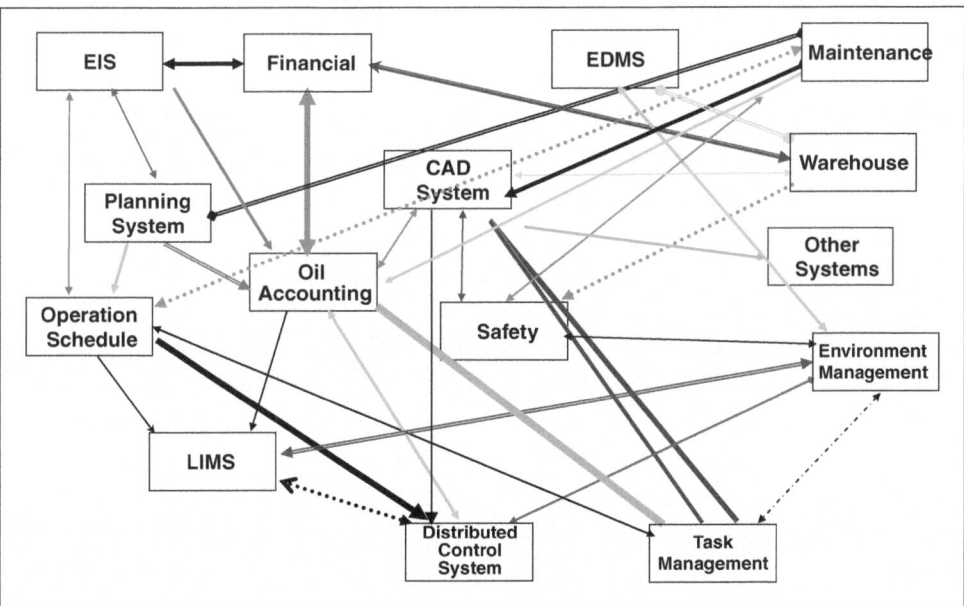

Figure 17-6 Computer Integrated Manufacturing Example

A second area for communication standards that has had significant impact on industrial automation is digital fieldbus. Fieldbus is a digital communication network for connecting intelligent plant devices into automation systems. Fisher-Rosemount (Emerson) introduced a hybrid analog/digital fieldbus called HART (Highway

Addressable Remote Transducer) that could communicate with devices and systems from different suppliers. The digital communication component of the HART protocol allows instrument information to be passed to a DCS to provide instrument health and similar analyses, but the measurement value is still transmitted via an analog signal. HART has been one of the most successful fieldbus implementations in industry.

A number of fully digital fieldbuses emerged following the success of HART. The two most prevalent in the process industries are the Foundation Fieldbus standard (controlled by the Fieldbus Foundation of Texas) and the Profibus standard (controlled by the Profibus Trade Organization out of Germany). A resulting fieldbus standard, Foundation Fieldbus, has become the primary standard for the industry. Standard digital communications between automation systems and field devices have provided a major step forward for automation systems architectures.

As human-machine interface (HMI) software and supervisory control and data acquisition (SCADA) software hit the industrial marketplace, one of the chief advantages of this software technology over other automation approaches was it worked with most of the automation systems and intelligent devices then available. This occurred through the inclusion of drivers (communication driver software) developed for each system or device to integrate with the HMI/SCADA software system. Wonderware developed an approach that capitalized on Microsoft's Dynamic Data Exchange (DDE) capability (which enabled communications between multiple applications operating in a Windows environment) to produce NetDDE, a networked version of DDE for the development of communication drivers.

HMI/SCADA software companies, systems integrators, and even industrial companies developed thousands of NetDDE drivers for intelligent systems and devices. Although DDE still worked in Windows environments, a newer approach developed by Microsoft, called Object Linking and Embedding (OLE) pushed it aside. As this took place, a standards committee started up to support an industrial version of OLE called OLE for Process Control (OPC). With the introduction of OPC, thousands of new intelligent device and system drivers all but replaced the older NetDDE drivers. These device drivers, whether NetDDE or OPC, are inexpensive and easy to implement, making the integration problem at the plant-floor level much less onerous than it had been.

As the intra-system technical integration of automation and business systems became easier and less expensive, the focus shifted to the development of integrated operations and business solutions using the available technologies. Systems integrators made the switch to become value-added solution suppliers, although they still called themselves systems integrators. But those who hung onto the traditional integration model have had difficulty surviving. Today, systems integrators are geared to identifying and solving operations and business problems through the effective use of both automation and business technology.

Standards have also moved more in the direction of application development rather than mere connectivity. For example, the ISA-88 batch series of standards focuses on the application of automation and information technologies to batch processing operations. ISA-95 is designed to identify key application and information constructs and issues when interoperating between plant operations and the business systems environment. MIMOSA (Machinery Information Management Open Systems Alliance) is a standard similar to ISA-95 but includes maintenance applications. Open O&M is a standard sponsored by the MIMOSA and OPC Foundations for combined operations and maintenance interoperation with business applications. This is representative of a very positive direction for industry, one in which the technology itself is no longer the driving force of industrial automation, rather the effective application of the technology is becoming a much more important issue.

The most recent trend in the development of technology environments that match the scope of industrial operations has been the move toward service-oriented architectures (SOAs), which were introduced in Chapter 16. According to the Organization for the Advancement of Structured Information Standards (OASIS), an SOA is "a paradigm for organizing and utilizing distributed capabilities that may be under the control of different ownership domains. It provides a uniform means to offer, discover, interact with and use capabilities to produce desired effects consistent with measurable preconditions and expectations." This is not too different from the initial definition of a computer operating system that enables the various functions within a single computer.

The difference is an SOA provides the distributed infrastructure and services that turn a distributed multiple computer system from multiple vendors into a homogeneous system across separate computers. SOAs can support different computer domains. For example, NetWeaver from SAP is an enterprise SOA (ESOA) designed to support a business enterprise domain, while ArchestrA from Invensys is an industrial SOA (ISOA) designed to support a real-time industrial domain. The development of SOAs is starting to significantly diminish the barriers to systems integration, providing common computer domains across multiple systems. Although these SOAs are still in their infancy, they will have a significant impact on industrial automation and business computing for decades to come.

The need to connect different intelligent systems and devices from different suppliers into a single interoperating environment has been felt ever since the introduction of the computer to industrial and business operations. The evolution of technology and approaches to support this need has been long and gradual. A very significant shift has taken place over the past few years from an intense focus on integrating the intelligent technology to an intense focus on solving business problems enabled by the technology. This is a major step forward for industrial automation because integration in and of itself adds no value; solving real business problems does.

Review Questions

1. Why is integration across automation system domains of value?

2. Why did automation systems integrators come into business in the first place?

3. How has the focus of automation systems integrators shifted over time, and why has it shifted the way it has?

4. What is MAP, and why was it significant?

5. What was the "vacuum cleaner approach," and why was it essentially ineffective?

6. Why did standardization activity shift from communication standards to application-based standards, such as ISA-88, ISA-95, MIMOSA, and Open O&M?

CHAPTER 18

Business Measurement and Intelligence Systems: Real-Time Knowledge

"A popular government without popular information, or the means of acquiring it, is but a prologue to a farce or a tragedy, or perhaps both."
- **James Madison** *wrote in 1822, five years after his presidency*

Knowledge is king and providing the right information at the right time to the right people is a necessity in today's automation environment.

With the evolution of business information systems and industrial automation systems, huge amounts of data have become available throughout the manufacturing enterprise. For years, management has wanted to use this data to perform business and operational functions more effectively. One of these functions should provide the proper information to every person in the organization to help them perform their jobs better. That is what business intelligence systems are all about. Business intelligence systems are computer-based software systems designed to develop and manage business measurements and the associated intelligence derived from those measurements.

Providing intelligence to workers in plants through process automation is certainly not new. Ever since the advent of automation systems, even before digital computer-based systems, one of the primary functions of automation technology was to provide a level of intelligence to operators, engineers, and maintenance workers. Panel boards and human-machine interfaces focused on this for decades. But most of the intelligence provided was operational, with limited business connotation. The challenge is now to provide high-quality operational and business intelligence to support the specific activities of each person in the plant, as well as those in a support function.

Traditionally, business intelligence was the domain of the business and accounting systems, and business executives and managers used it to determine how effectively the overall business was operating. For decades there was little interest in sharing this information with the folks responsible for plant operations, at least in any effective format. The primary reports the operations personnel have traditionally received from ERP software in the business information systems have been monthly variance reports (Figure 18-1) that summarize the cost per unit of product produced against the expected cost per unit of product, usually referred to as standard cost. If the actual cost per unit was less than the standard cost, the production operation did a "good" job. If the actual cost was greater than standard cost, the production operation did not do a good job. Although there is not much useful information content in variance reports, they were the primary source of business intelligence to production operations for decades.

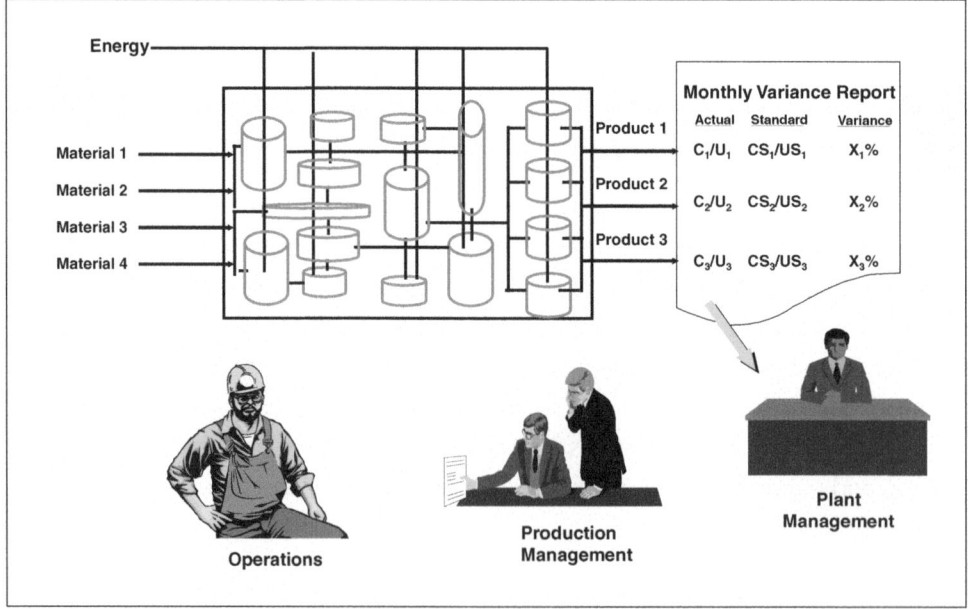

Figure 18-1 Monthly Variance Reporting

As the amount of data available in industrial operations increased, a drive to provide more effective business intelligence throughout the organization developed. One of the best known initiatives in the drive for better business intelligence came in the 1980s from Dr. Robert Kaplan and Dr. David Norton of Harvard University when they introduced the concept of the Balanced Scorecard (BSC). The idea behind the initial forms of BSC was to provide more balanced information to managers to help them manage their operations more effectively. Rather than just using financial information, Kaplan and Norton suggested balancing the financials with additional information relevant to the management of the operation, such as customer satisfaction, innovation, learning and internal perspectives of the business. This was certainly a huge step forward, but the target was operations management, not operators and maintenance personnel. Every person in a production operation who takes actions that may have an impact on the performance of the business also needs a level of business intelligence.

Executives in industrial businesses have long raised concerns over the lack of timely business and financial visibility into their manufacturing operations. Most executives do not really know whether the production operations have been performing well or poorly until a number of days after the end of the month, due to the monthly nature of the financial measurement system. In the changing manufacturing environment, where being nimble and agile are key, this is unacceptable. Ten years ago most energy contracts between energy suppliers and industrial businesses were on a fixed price for energy over an extended period of time, up to a year. This essentially relegated energy cost to a constant over that period. Today, the cost of purchased energy changes multiple times in a single day. Business is starting to become much more real-time. Business intelligence must be just as dynamic. It must be real-time.

The model in Figure 18-2 came from a number of executives to capture their perspective on how automation and business systems enable them to manage their industrial operations. The top section of the model represents the executives' perspectives on how they execute their job. They basically view the business functions as falling into two categories: operating the business, which is the domain of the Chief Operating Officer, and measuring the business, which is the domain of the Chief Financial Officer. Under each of these basic business function categories, the executives developed a simple model of how automation and information technologies combine to provide support for each.

Under the COO is a three-level stack of functionality that is essentially partitioned by the timeframe in which the functions need to operate to be effective. The lowest level represents all of the functions that must execute in real time to be effective, such as process control, logic control, and operator interfacing. The second level in the stack represents the functions that need to operate on more of a daily schedule to be effective, such as production planning and production scheduling. The top level represents functions that must operate on more of a monthly basis, such as supply chain optimization and customer relationship management.

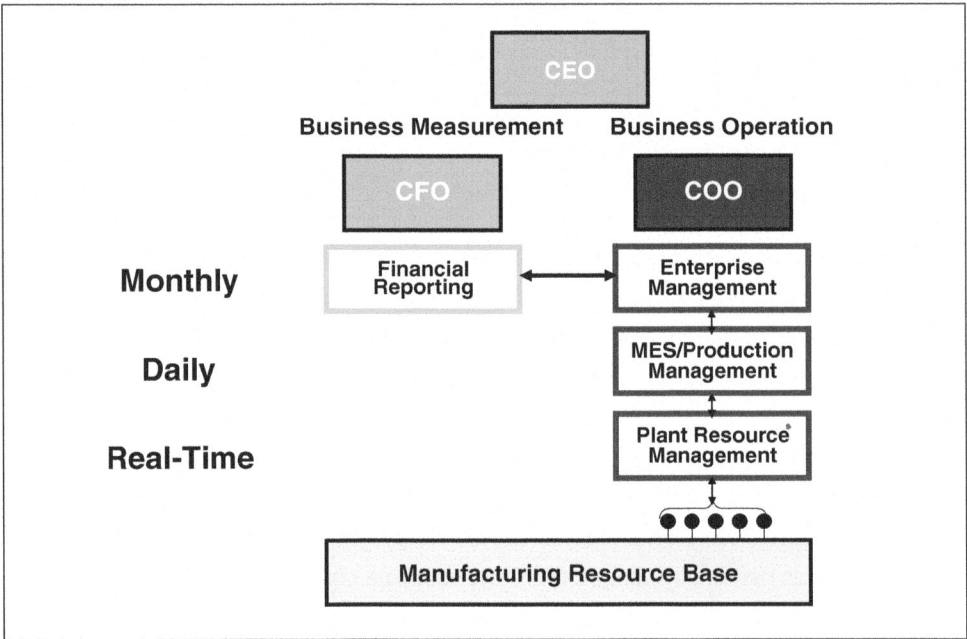

Figure 18-2 The Accounting Gap

Although this model is quite simple, it points out a huge need in industrial businesses typically not being met today. The left-hand side of the model represents the functions commonly available today for measuring the business. Notice the left-hand side only has the topmost level, "Financial Reporting." This is because most industrial operations only do ERP-based (enterprise resource planning) financial reporting once a month as the primary measure of the business. The problem with this is any functions in the lower two functional levels on the operations side are not measurable by the financial measurement system. Before anyone can have an effective business intelligence system, the functional gap at the lower two levels of the business measurement stack must be addressed.

One interesting sidelight to this problem is that in some industrial companies, the operations team tried to fill the measurement gap by developing operational measures, called key performance indicators (KPIs), to measure the lower two functional levels of the operating stack. In many respects the KPIs filled the void, but they had little credibility with the financial and accounting teams. Although developing KPIs is a useful activity, without the corresponding financial metrics the measurement system and the resulting business intelligence system are severely lacking.

The key issue remains how to fill in the lower level gaps in the business measurement model. Financial professionals have tried to get the accounting systems down to daily from monthly by the development of Activity-Based Costing (ABC) systems, but the ABC systems still, for the most part, employ a top-down approach to accounting. For the financial measurement system to meet the requirements, it must provide real-time data right down to the plant floor, which requires a real-time database that can be used as inputs to the accounting calculations. The good news is such a database exists in the form of the sensors in the plant that measure flow, level, temperature, pressure, composition, speed, and many other variables. Although this is not a financial database, engineers can use this data to model the financial equations right in the control system. This can be accomplished by having the accounting team in industrial plants develop the equations for the required accounting measures for each process unit. These equations are provided to the engineering team, who has expertise at determining which process measurements can be used to resolve models of the equations running right in the control systems. The result is a bottom-up, real-time accounting approach that can flow right into the ERP financials, as shown in Figure 18-3.

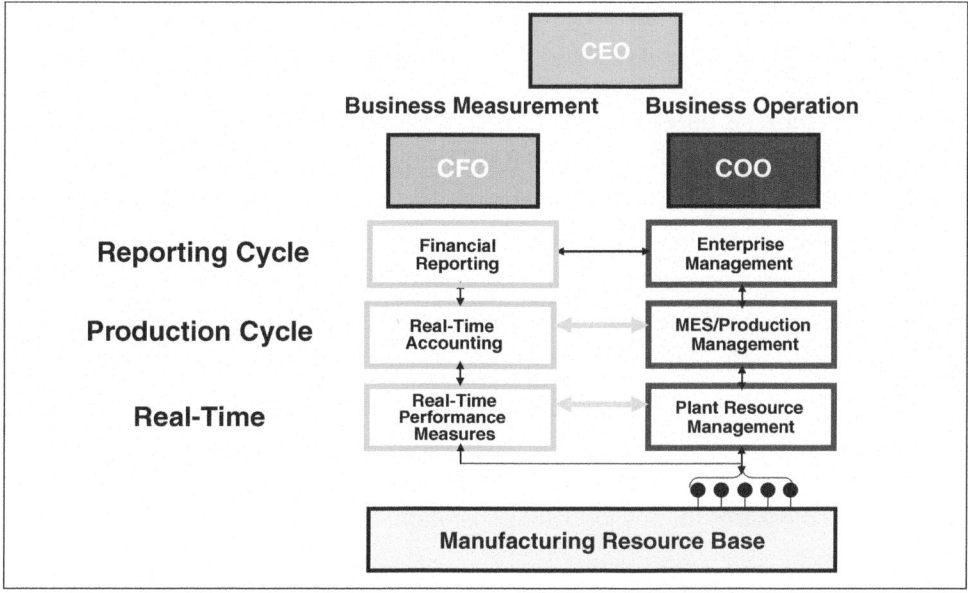

Figure 18-3 Performance Management

The "Real-Time Performance Measures" correspond to the "Plant Resource Management" level on the operations side of the model and provide effective financial measures in the real-time domain. The "Real-Time Accounting" function on the financial side corresponds with the "MES/Production Management" functions on the operations side of the model, providing effective measures of those functions. Adding the two lower levels to the financial side of the model fills a traditional information gap but does not provide all the information necessary for an effective business intelligence system. Nonetheless, it certainly is a step in the right direction.

For employees to get the max out of industrial business intelligence, it has to occur in a time frame that enables each employee to take effective action, and it has to be in the proper context for each area of responsibility. In production operations, the time frame that is most effective at all levels in the organization action is real time. Manufacturing and production are inherently real-time functions, and therefore they must be monitored and managed in real time for maximum impact.

Also, to be actionable and effective, the business intelligence provided to each person in the organization must specifically focus on that person's domain of responsibility. Manufacturing and production processes are inherently complex, which is why most of these processes are partitioned by both operations and maintenance. Providing a plant-wide perspective to an operator who only has responsibility for a section of the plant can be misleading and frustrating and can sub-optimize the actions that operator takes. The only way to meet the real-time and contextualized requirements for business intelligence is to build the business intelligence system by starting at the plant floor and working up through the organization.

The good news is developing bottom-up business intelligence systems does not require any new hardware or software. It only requires using these technologies a little differently. One of the first requirements is to convert the operational KPIs from daily measures to real-time measures. This can happen in a similar manner as developing real-time accounting measures. As has been mentioned, every industrial operation has a large number of instruments installed throughout the process that measure all kinds of physical and chemical variables in real time. Using these measures as the data source for the calculation of the KPIs right in the control system environment provides real-time KPIs. Real-time KPIs can combine with the real-time accounting measures to provide a fairly complete set of real-time performance measures for any industrial operation.

If these real-time KPIs and real-time accounting measures come online for each process unit or work cell in the operation, they can be mapped to each employee's domain of responsibility. The result may be a fairly large set of performance measures that define the performance of each person in the plant.

It is quite difficult for any person to respond effectively to a large number of performance measures competing for attention. The most effective way to limit the information must be put into the context of the current manufacturing or production strategy of the plant as detailed in Chapter 19. Typically, this contextualizing to strategy does not create new measures of performance; instead, it prioritizes the existing set of financial and operating performance measures (also referred to as metrics) so each person in the operation can understand which of their performance measures is most important to the operation right now, which is second and so on. This strategy essentially provides a lens through which operational and financial performance measures can be focused and prioritized.

These real-time, prioritized performance measures are dynamic performance measures (DPM). Studies have shown that people working in real time can effectively deal with up to four competing measures. Therefore, business intelligence dashboards presenting the highest priority four measures relevant to each plant person provide the most effective plant-level business intelligence system (Figure 18-4).

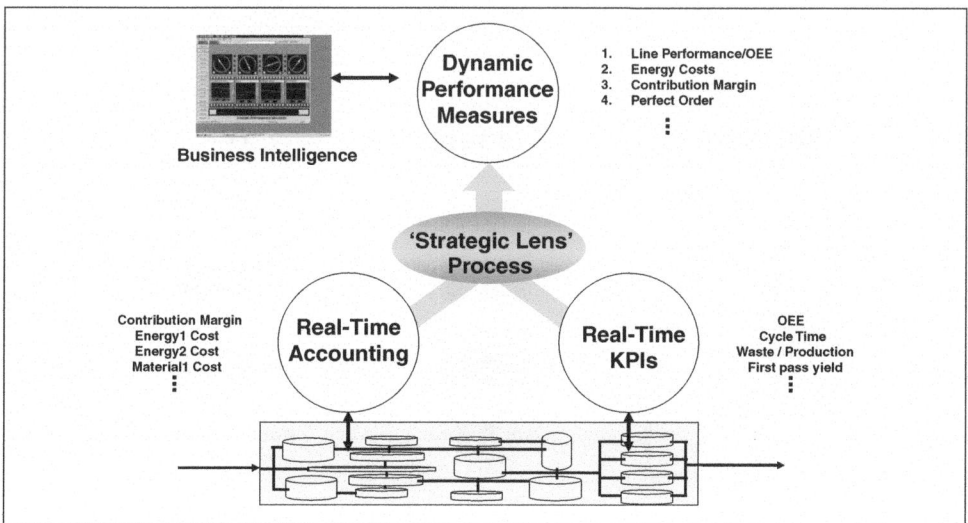

Figure 18-4 Strategic Business Operational Measurement System

Once the real-time plant business intelligence system is in place, the corporation can begin the development and implementation of the remainder of the business intelligence system by working up through the organization, developing composite bottom-up measures for each manager in the structure. These measures will certainly need support by additional information important to each job function, such as current commodity prices, production schedules, sales demand or whatever other information may be relevant to the responsibilities of each job. Using the aforementioned balanced scorecards or another proven approach for the display of real-time information can provide an effective and familiar means of communicating performance information at the management levels. Figure 18-5 presents a simplified model for bottom-up, contextualized industrial business intelligence system.

Real-time business intelligence can truly empower all levels of industrial operations to be able to perform their tasks more effectively and drive significant performance improvements throughout the operation. For decades, industrial companies have been searching for ways to replace people with technology. Today, using technology to make plant and non-plant employees more effective is becoming more important than it has ever been. People can perform brilliantly if they have the timely, complete, and accurate information they need to make informed decisions in the right time frame. Plant-floor to enterprise business intelligence systems can provide this information in real time—and the results can be significant.

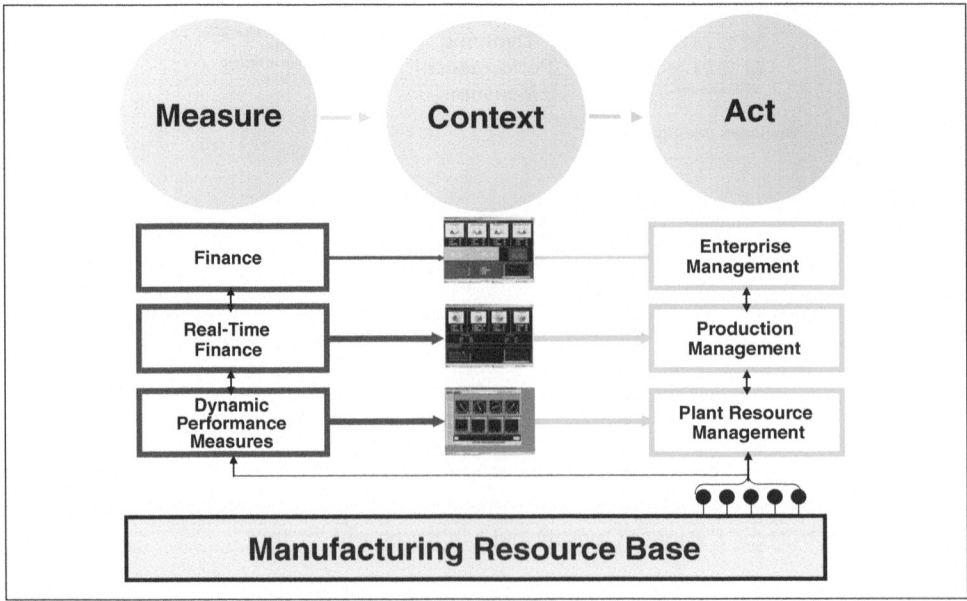

Figure 18-5 Enterprise Business Intelligence

Review Questions

1. Why are monthly variance reports developed in the financial management system inadequate for providing effective performance intelligence to production operations?

2. How did the Balanced Scorecard approach advance the state of the art with respect to business intelligence?

3. What is ABC?

4. What are dynamic performance measures?

5. Why is a multilevel business intelligence approach effective?

6. Why is it important to provide business intelligence to frontline workers in real time?

CHAPTER 19

Operations Business Excellence: A New Frontier

Quite a while ago, in the 60s as a matter of fact, there was a television show called *Star Trek*. At the beginning of every episode the show's commander, James T. Kirk would always say, "Space... the Final Frontier. These are the voyages of the starship *Enterprise*. Its five-year mission: to explore strange new worlds, to seek out new life and new civilizations, to boldly go where no man has gone before."

When you talk about "where no man has gone before," take a look over the past few decades. There have been significant technological developments, made under the banner of operations excellence (OE), that can improve process optimization and asset management within industrial plants, but with the change in perspective that comes from operations business excellence (OBE), manufacturers are starting to reach new thresholds of performance improvement. Technologies developed years ago to solve basic plant issues, such as basic process control technologies, may now work in different ways to control the business of the plants rather than just the process loops.

Let us explain.

Operations Excellence

The phrase "operations excellence" has become a catchall phrase for the technologies, methodologies, and initiatives deployed to generate the most value possible from plant assets. OE, therefore, is really not very well defined, but initiatives under this banner have provided some very important and interesting steps forward in the drive for performance improvement in industrial operations.

To a large extent, OE has been associated with "doing things right" within the manufacturing operation. To really do things right requires the user to be able to measure what "right" really is. Over the past few decades, the measures of OE have been the key performance indicators (KPI) of the operations. These KPIs are often developed by the operations management and engineering teams and focus on the things these teams determine must be done effectively to move toward excellence. Dozens of different KPIs have been developed across industry to try to get a handle on how each operation is doing and to drive continuous operations improvement. Figure 19-1 lists a number of the KPIs presented in the ISA-95 specification. Although these KPIs may have been set up to measure the critical items each operation needs to focus on to drive improvement, there has been very little consistency across the industry as to which KPIs should be measured and how they should be measured. There have also been so many KPIs implemented at industrial companies that it has often become very difficult to determine how the organization is really doing and whether it is really improving. Sometimes too much data is more constraining than not enough.

Actual production rate as a % of the maximum	Actual inventory turns	Order line fill rate
Actual vs planned volume	Annual work -in process (WIP) turns	%error in cases shipped
Average machine availability rate	Customer order cycle time in days	% error in orders shipped
Average machine uptime	Finished goods inventory turns	% of orders shipped complete and ontime
First product, first pass quality	Inventory accuracy	% of sales orders delivered on time
Hours lost due to equipment downtime	Inventory reliability: the items filled on	Pilfering reduction
Major component first -pass yield	first try per total line items ordered	Stock turns per year
Manufacturing cycle time for a typical product	Lines shipped per person hour	Pallets shipped per person per hour
No of process changes per operation due to errors	Rework and repair hours % of direct hours	% error in lines shipped
% error in yield projections	Scrap and rework as % of sales	% of orders expedited
% increase in Output per employee	Scrap and rework % reduction	% of sales order line items not fulfilled - stock outs
% of assembly steps automated	Standard order-to-shipment lead time	% of supplier orders delivered on time
% of lots or jobs expedited by bumping other	Time line is down due to sub-assembly shortage	Pick-to-ship cycle time for customer orders
lots or jobs from schedule	Time required to incorporate engineering changes	Raw material inventory turns
% of operators with expired certifications	Units produced per square foot	Vendor lead times
% tools that fail certification	Warranty effort reduction	Labor hours spent on preventive maintenance
% reduction in component lot sizes	Warranty repair costs as a % of sales	Maintenance cost as a % of equipment cost
% reduction in manufacturing cycle time	Yield improvement	Maintenance cost per output unit
% unplanned overtime	% error in reliability projections	Number of unscheduled maintenance calls
Production and test equipment setup time	% of lots going directly to stock	% of equipment maintained on schedule
Production schedules met (percentage)	% of product that meets customer expectations	Unplanned machine downtime as a
Productivity: units per labor hour	% of quality assurance personnel to total personnel	% of scheduled run time
Reject rate reduction	% of quality engineers to all engineers	Stock turns per year
Actual inventory turns	Receiving inspection cycle time	Annual lines shipped per SKU
Annual work -in process (WIP) turns	Time required for corrective action	Cases per hour
Customer order cycle time in days	Time to answer customer complaints	Dock-to-stock cycle time
Finished goods inventory turns	Time to correct a problem	Gross inventory as a % of sales dollars
Inventory accuracy	Variations between inspectors	Inventory carrying cost
Inventory reliability: the items filled on	Lines shipped per person hour	Line items processed per employee/hour
first try per total line items ordered		Order fill rate

Figure 19-1 ISA-95 Listed KPIs

Industry continued to search for a simple and universal answer through a simple and universal performance measure that could be used in the same way in multiple different industrial operations and could provide a benchmark for performance excellence. The result of this search was operations equipment excellence (OEE), which appeared to provide a single performance measure for any industrial operation and could therefore provide a benchmark. OEE is defined to be the product of asset availability, product rate, and quality over a given period of time, expressed as a percentage. With the common acceptance of OEE as the primary measure of OE, industry finally seemed to have the long anticipated performance measure of excellence. Industrial organizations soon began to build OE initiatives around the OEE measure.

The truth is many of these organizations have realized some significant improvements with the additional attention given to the operation. Although these initiatives have been very promising, a composite performance measure such as OEE can also present deceiving results if it is not carefully implemented and managed. Experience has demonstrated how the value of OEE that represents good operation can vary significantly across different manufacturing equipment types due to the nature of the operation performed in the equipment. This means there is no universal understanding of which value of OEE actually represents operational excellence and no standard way of calculating it across organizations. This limits its value as a universal benchmarking tool.

Another issue with OEE is each of the three factors, asset availability, product rate and quality, can have inverse effects on the other two. In other words, to push product rate higher may lead to reductions in availability and quality. The overall value of OEE may not change, but the inverse nature of the factors of OEE may drive organizational conflict.

Finally, OEE, as a product of three factors, can actually hide a problem in the operation and present a good result while one of the three factors is actually on the verge of dropping off at a significant rate. None of these issues means that OEE is not a good and effective measure. It is. But when using OEE, you should employ caution.

OEE can be a very good and effective management reporting measure, but it tends to be an ineffective actionable measure. That is, OEE does provide a fairly solid indicator of improvement within and across operations and is as good a KPI as any in providing a reporting measurement of this type. However, it is difficult for operations, maintenance, or engineering to know what action to take when OEE starts to decline. For example, a decline in the OEE value may be the result of an equipment availability problem, or perhaps a product rate problem, or perhaps a product quality problem, but the value of OEE does not provide any insight on which of these it really is. What makes OEE valuable as a reporting measure actually causes it to lose value to the team that has to take action. OEE is necessary, but not sufficient.

Figure 19-2 Operations Business Excellence

The question is, what measures, in addition to OEE, are required to set the stage for true OE? The most effective answer to this question requires a little more analysis of what OE should include in order to drive results in a manner that industrial executives would like. OE is "doing things right," but is doing things right enough to get the results from production operations that business executives need? Many executives have claimed it is not. Not only do industrial companies need to "do things right," but they also need to *"do the right things."* Doing the right things has been referred to as strategic excellence (SE). SE and OE together are important, but publicly traded companies also have to generate the right value from their production assets, a value referred to as business excellence (BE). The application of any of these three measures of excellence, without consideration of the other two, will provide suboptimal results. There is a general recognition across industry where all three aspects of excellence must be combined into a single model for driving excellence in industrial operations (Figure 19-2). This combined approach is sometimes referred to as operations business excellence (OBE).

Developing Operations Business Excellence

In order to develop an OBE environment, a common set of performance measures must first be developed that represents a composite of these three traditional measures of excellence. The first issue with the development of a composite set of metrics is strategic, operations, and business excellence are each measured in different time frames. Operations excellence is traditionally measured using KPIs on a daily basis. Business excellence is measured through the plant accounting systems, which traditionally has been done on a monthly basis, and strategic excellence has been measured over much longer time frames, with quarterly being about the shortest. Perhaps these three different time periods made sense in an age when reasonable external business drivers and strategic market drivers changed on no more frequent a basis than monthly and quarterly. But the business and strategic environment today is much more dynamic than it ever has been, with business drivers and market drivers sometimes changing multiple times in a single day. Today's dynamic strategic business environment necessitates operations, business, and strategic performance measures all in real time.

The second issue with the development of a composite set of performance measures is the business space over which the measures have traditionally been developed. Operations measures are often made right down to the individual process unit or work cell. Business measures are typically made for an entire plant and up into the enterprise. And strategic measures are typically made for the overall company. Implicit in the three different space frames for the development of performance measures is no strategic decisions are modeled at a level lower than the overall enterprise and no business decisions are made at a level lower than the plant's management.

Neither of these should be the case. Frontline employees are called on to make hundreds of important decisions each day. Some of those decisions impact the business and some impact strategic execution. If the measures of business and strategic performance are not made available down to the front line, the decisions will be made blindly, which cannot be good for the performance of the corporation. Therefore, all three types of performance measurements, operational, business, and strategic, must be made in real time right down to the process units or work cells.

Developing a unified performance measurement system in real time down to the plant floor requires a real-time data source down at the plant floor. Fortunately, as we have seen in earlier chapters, such a data source exists in the form of the hundreds of sensors installed throughout industrial plants to measure flows, levels, temperatures, pressures, speeds, compositions, and the like. This extensive data source can provide real-time data to executing algorithms in the control software that can model the strategic, operations, and business performance measures of the process units or work cells in real time. Once these base performance measures are developed, they can be aggregated from a time perspective, resulting in hourly, shift, daily, weekly, and monthly summaries of performance by using a standard process historian. They can also be aggregated from a space perspective from units, to areas or trains, to plant sections, to plants, and to the entire enterprise merely by combining the performance measure models appropriately for each node. The result is an OBE performance measurement approach, a comprehensive bottom to top performance measurement system offering complete organizational alignment.

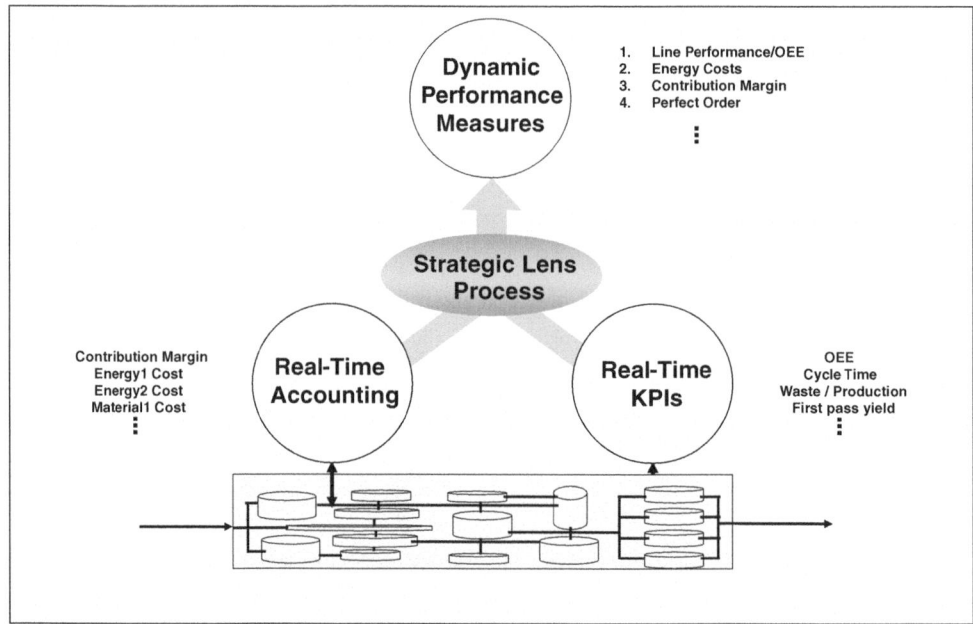

Figure 19-3 Business Operations Measurement System

The development of such an OBE performance measurement approach is actually simpler than it seems. Developing real-time KPIs and real-time accounting measures is fairly straightforward because the operational (KPI) and business (accounting) measures are based on equations fairly well understood and documented. Once the equations are developed, modeling them from real-time process sensor-based data is an engineering exercise. The difficulty often encountered is the development of the strategic performance measures. As was pointed out in the business intelligence chapter, strategic performance measures are really not additional measures, rather they are the combined KPIs and accounting measures prioritized to support strategy. In other words, the production strategy serves as a lens through which the basic accounting and operational measures are ordered according to strategy (Figure 19-3).

The combined strategic, business, and operations measures, commonly referred to as dynamic performance measures (DPM), provide the basis for the implementation of a true OBE environment.

One of the challenges with this approach to the development of an OBE performance measurement system is the deployment of an effective strategy analysis process that enables the development of the "strategic lens" through which the composite performance measures can be prioritized. A very straightforward strategy analysis (decomposition) process, which can be an effective aid in developing the strategic lens, was developed by Dr. Thomas Vollmann (Figure 19-4).

The Vollmann decomposition approach is a top-down process that utilizes a very simple construct, a strategic triangle, to analyze strategy. The strategic triangle has three vertices with strategy at the top, action steps at the left-hand lower vertex and performance measures at the right-hand lower vertex. The process starts with defining the current production strategy within the overall corporate strategy as a set of actionable strategic production objectives. An action plan is then developed to define the execution steps for each of the strategic objective. Each step in the action plan should be measurable, which is how the strategic performance measures are identified. The output of this process provides a mapping of the operations and accounting measures to strategy. The prioritized measures can then be decomposed down through the process areas all the way to the process units, providing a set of strategic performance measures for each process unit in the plant.

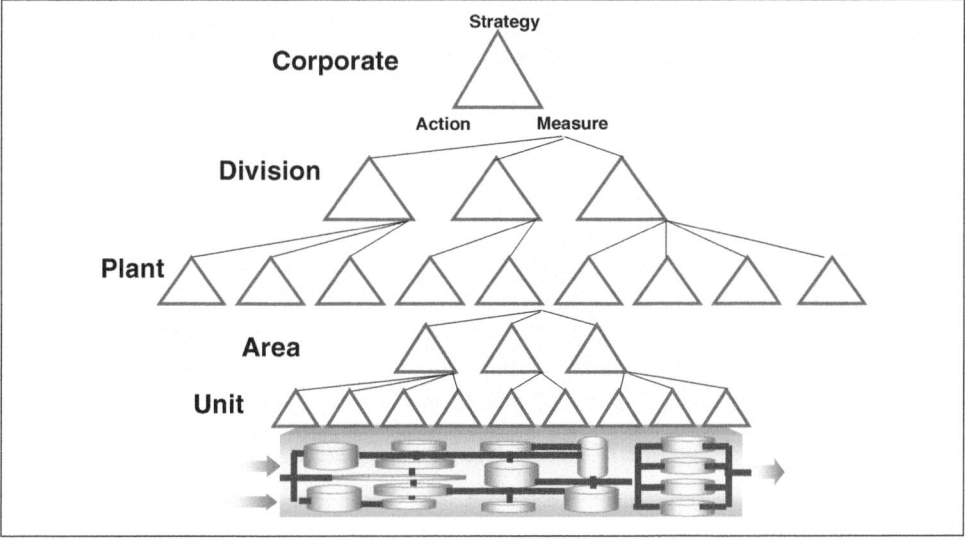

Figure 19-4 Vollmann Strategy Decomposition

Once the OBE performance measurement system is established, the next step in realizing OBE is providing all personnel in the production operation with the subset of strategic performance measures that apply to their specific domains of responsibility. This can be accomplished by developing real-time dashboards and scorecards as prescribed in the business intelligence chapter of this book.

The third issue associated with the development of a composite set of metrics of an OBE system is a specific focus on the performance improvements that can be generated through the application of advanced technologies, such as advanced process control or advanced asset management approaches. Traditionally, maintenance and operations have been managed as somewhat independent functions in industrial plants. This really tends to be a limiting factor on performance improvement because the operations and maintenance teams are working on the same set of assets in the plant.

One of the basic concepts of OBE that is quite different from traditional OE, which focuses on each function independently, is that collaboration between critical functions such as these can generate incremental value from plant assets. One problem is the traditional approach to judging the performance of maintenance and operations actually puts these two in conflict with each other.

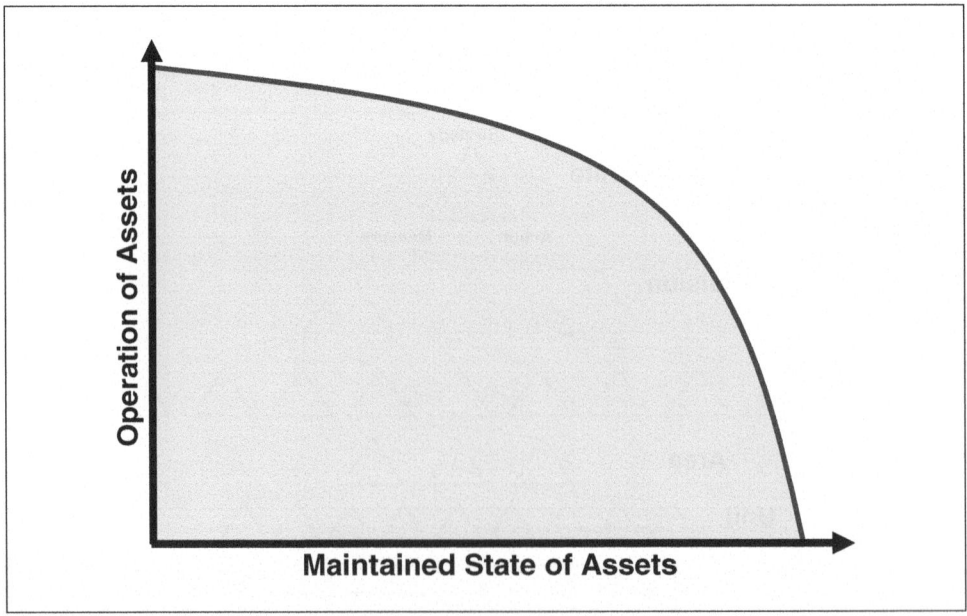

Figure 19-5 Inverse Relationship Between Maintenance and Operations Metrics

If, for example, the maintenance team is measured on the current maintained state of the plant equipment and the operations team is measured on the current operation of the equipment to optimize production output, it becomes clear these measures represent inverse functions (Figure 19-5). Driving more output through the equipment through process optimization initiatives will typically cause a decline in the maintained state, in spite of the well-understood risk of unplanned downtime due to equipment failure. Conversely, improving the maintained state of the equipment through asset management initiatives will typically necessitate shutting down some of the equipment, which will adversely impact production output. It is no wonder that these two teams have difficulty cooperating.

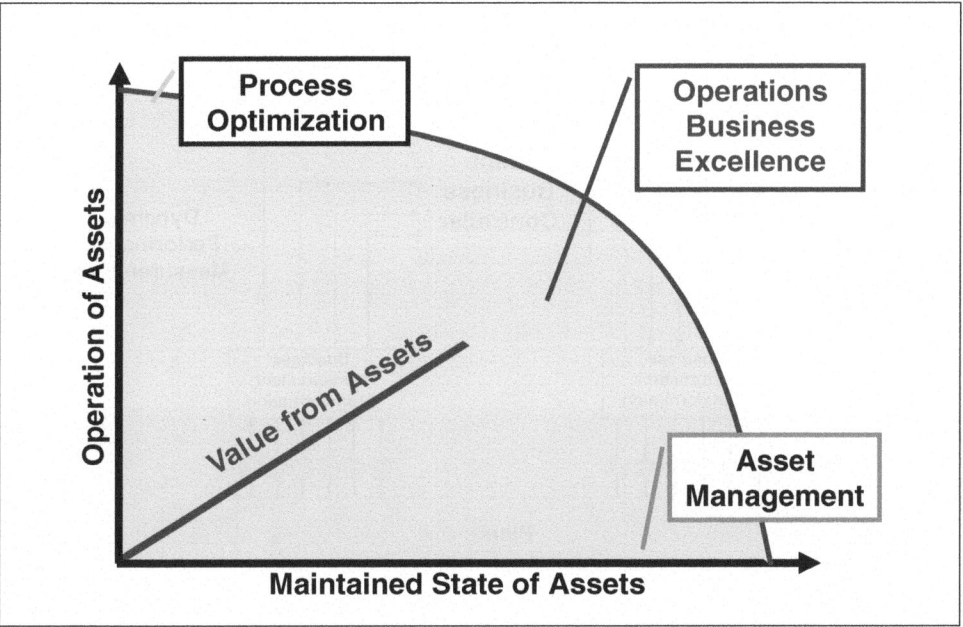

Figure 19-6 The Asset Value Vector

The key to effective OBE is to focus conflicting organizations on common measures of performance. In the case of operations and maintenance, the common measures are the strategic performance measures of the business. These strategic performance measures represent the value being generated from the assets (Figure 19-6). If both the operations and maintenance teams focus on the value being generated from the assets, significant improvements in the business value being generated from the industrial assets will be realized.

These same value criteria can be used to judge the value of the application of any advanced process optimization or asset management approach for improving the value generated from the production operation. Industrial plants can focus on the tools and initiatives that drive improvement and abandon those that do not. This approach enables industrial operations teams to go well beyond traditional OE to true OBE and become the highest value-adding teams in industrial organizations.

With the development of the strategic performance measures, control theory is starting to be applied to these higher level measures rather than to just the basic process measures. The looked-for result is closed loop business control (Figure 19-7).

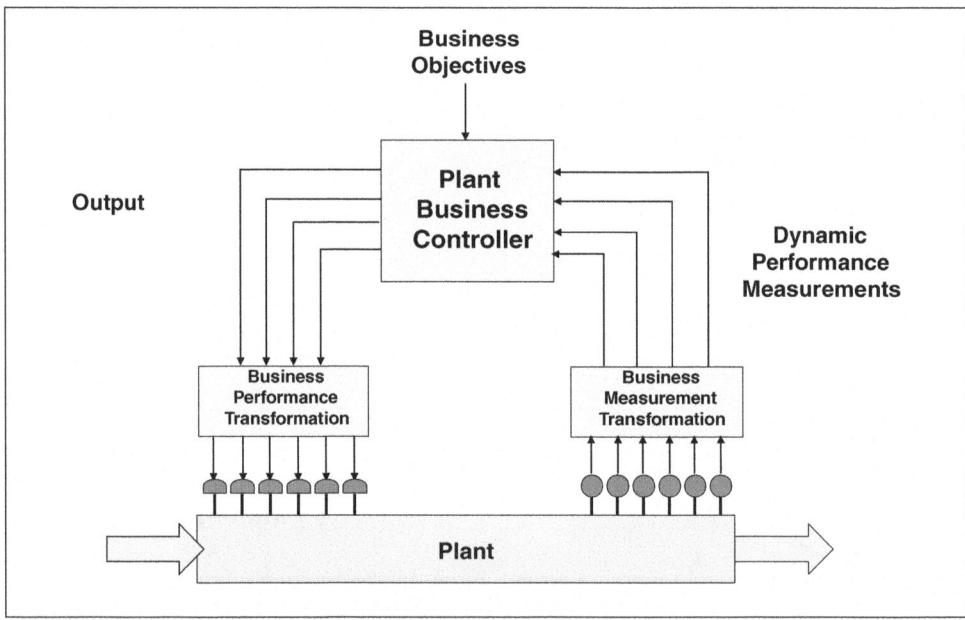

Figure 19-7 Closed Loop Business Control

The application of advanced engineering techniques to the control and management of industrial businesses is only starting to be seriously investigated, but already the results are very promising. Operations business excellence is not achieved by moving away from the competencies that have been required to make complex industrial plants operate as effectively as they do, rather it is achieved by applying these competencies at the business levels of the organizations. Operations business excellence has been considered to be an unobtainable ideal, but as automation technology advances it is starting to become a reality.

Review Questions

1. What does KPI stand for and how are they commonly used in industrial operations?

2. Why are KPIs as they are implemented today necessary but not sufficient?

3. What are the three fundamental components of an effective operations business excellence approach?

4. Why do dynamic performance measures have to be prioritized according to the production strategy?

5. What is a composite KPI?

6. Why are composite KPIs useful and what are their limitations?

7. Why is closed loop business control a valuable concept?

CHAPTER 20

Enterprise Control Systems: Grabbing the Technology Edge

L et's face it, in today's automation environment; it is all about communicating up, down, and throughout the enterprise. That communication allows companies to squeeze as much profit as they can out of disparate systems about as compatible as the Hatfields and the McCoys.

Just think about it, among safety systems, DCSs, PLCs, MES, advanced control and optimization, ERP, simulation, SCADA, and LIMS there are more than enough different types of automation and information systems not designed to interoperate. When you really look at it, these software and system offerings evolved independently and having them work together just was not even considered. That all ended up creating one of the biggest buzz phrases in the early twenty-first century, "islands of automation." These islands became costly technological barriers to industrial companies in search of solutions that might need to span more than one of these domains. Although there was a general feeling that cross-domain solutions would add value, few industrial companies dared to spend the money to build bridges to these islands.

The need became apparent for new technological approaches to pull all of these systems and software islands together into a single system that could connect the islands and cover entire plants and even entire industrial enterprises in a cost-effective manner.

These islands fell into three levels of functionality (Figure 20-1): automation, manufacturing execution systems (MES), and business. Suppliers providing products to all three levels claimed to have "open" systems or software. But the concept of "openness" seemed to vary significantly from level to level. At the automation system level, most suppliers were trying to be as open as reasonable within the higher priorities of safety, security, and environmental protection. The problem was with openness not being one of the top three priorities, these systems were not very open. A similar phenomenon was evident at the business level in which financial reporting, data integrity, and business security were higher priorities than openness. Interestingly, the MES level had exactly the opposite characteristic. Openness was a top priority for this software because this software had to work with either automation systems or business systems or both to execute an intended function. MES software typically evolved with huge numbers of point-to-point connections to both automation and business systems. Although the MES level was more open than the other two, the state of interoperation of critical information and automation assets was clearly a barrier to the implementation of many business solutions.

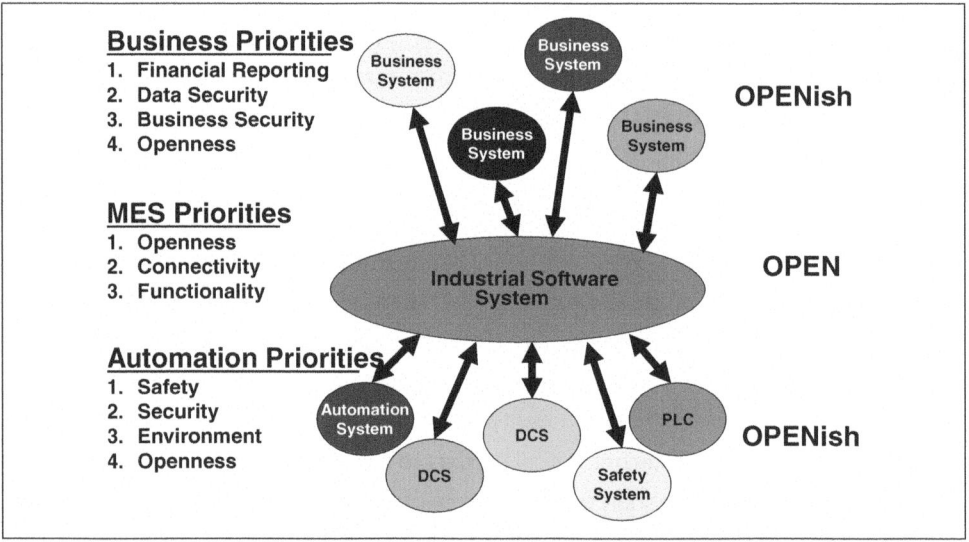

Figure 20-1 Prior to Enterprise Control Systems

As we have previously seen, the technological development that overcame the interoperation problem was service-oriented architectures (SOAs). According to the Organization for the Advancement of Structured Information Standards (OASIS), an SOA is "a paradigm for organizing and utilizing distributed capabilities that may be under the control of different ownership domains. It provides a uniform means to offer, discover, interact with and use capabilities to produce desired effects consistent with measurable preconditions and expectations." In a sense, an SOA is like having a common distributed operating system that goes beyond any single computer platform and provides services that make this highly distributed non-homogeneous environment work as though it were a single computer.

The first generally available SOAs came from IBM, SAP, and other major suppliers to the business information management markets. They pulled disparate software either from within their organizations or from the outside marketplace into a common computing domain. SAP provided a good example of an enterprise SOA (ESOA), NetWeaver, which pulled different enterprise level software applications into a suite in which all the applications could interoperate.

These enterprise SOAs provided a major enhancement at the business software level, and on paper they should have provided similar capability at the industrial level, within the plants. But as they say in any plant, that is the difference between what you learn in school and reality. The problem is the services required to make industrial plants operate effectively, safely, and in an environmentally responsible manner are very different than those required to make business software work together. That is why a new class of SOAs had to be developed to meet the needs of industrial environments, including real-time response, security, object management, and common name space services to name a few.

This new class of architecture is referred to as industrial service-oriented architecture (ISOA) or sometimes manufacturing service-oriented architecture (Figure 20-2). ISOAs have some of the same services found in enterprise service-oriented architectures (ESOAs) but also have a number of additional services designed for industrial operations. ISOA technology can be used to pull together the disparate industrial systems and software to make the resulting set of components appear to work as a single system, almost as though they were designed from inception to operate in that manner.

Business and automation systems and software are essential to industrial operations. Since different SOAs are required for business and automation systems, to develop enterprise-wide coverage requires an ESOA and an ISOA and requires these two classes of architecture to interoperate (Figure 20-3). Some SOA providers are offering certification programs that ensure either third-party applications or third-party SOAs can interoperate with their SOA. Certifications of this kind provide industrial companies a level of comfort and the opportunity to implement a multi-level architecture that covers all of their business and automation requirements.

Since some traditional MES software can interoperate with either an ESOA or an ISOA and this software typically brings with it all of the point-to-point connectivity it previously supported, the combination of these new SOAs and this MES software can create an architectural environment that is state-of-the-art and can interoperate with automation systems and software installed as far back as 30 years ago. This combination is very important because, unlike business systems, automation systems tend to have very long effective lifecycles. Without this connectivity into traditional systems and software the power of ISOAs may not have been realized for years.

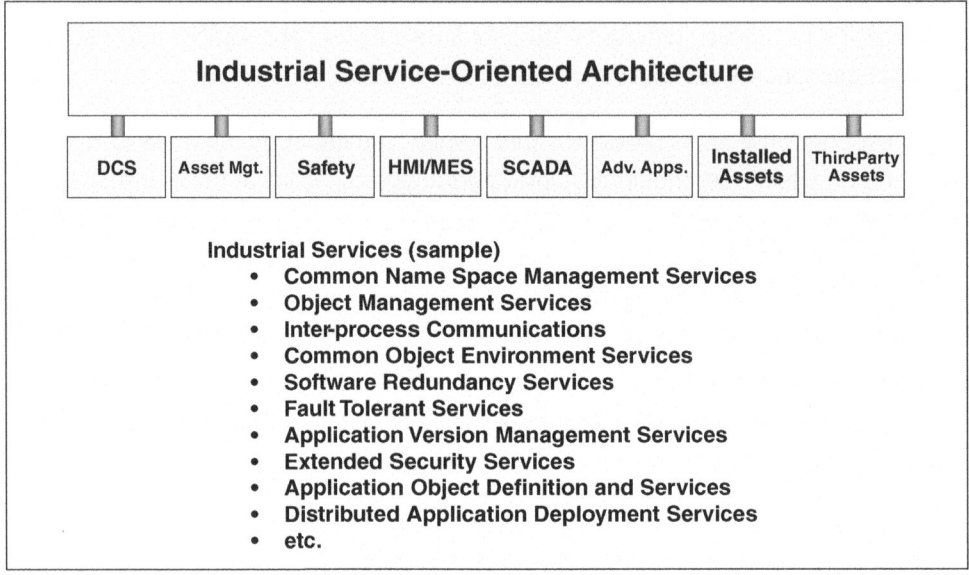

Figure 20-2 Industrial Service-Oriented Architecture

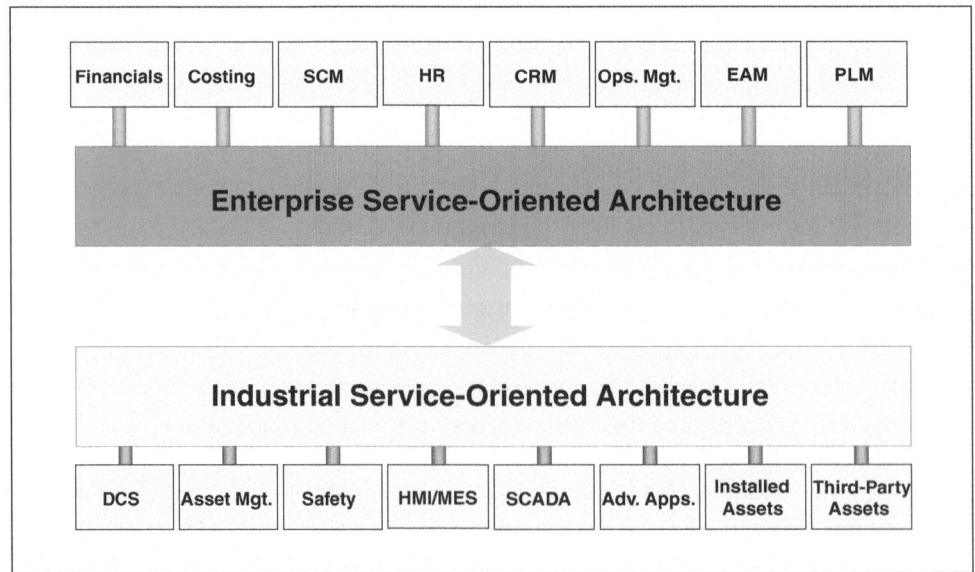

Figure 20-3 Enterprise Systems Convergence

Once an industrial company has selected and installed an ESOA at the business level and an ISOA at the industrial level, the combined architecture will allow for the islands of automation and information to absorb into a much broader compute space. An industrial company can start with the software and systems they have installed over the years and build a new computing architecture that behaves essentially as a single system across the breadth and depth of the industrial enterprise (Figure 20-4). Installed systems that operated in a totally independent manner in the past automatically interoperate in a single unified enterprise and automation compute space.

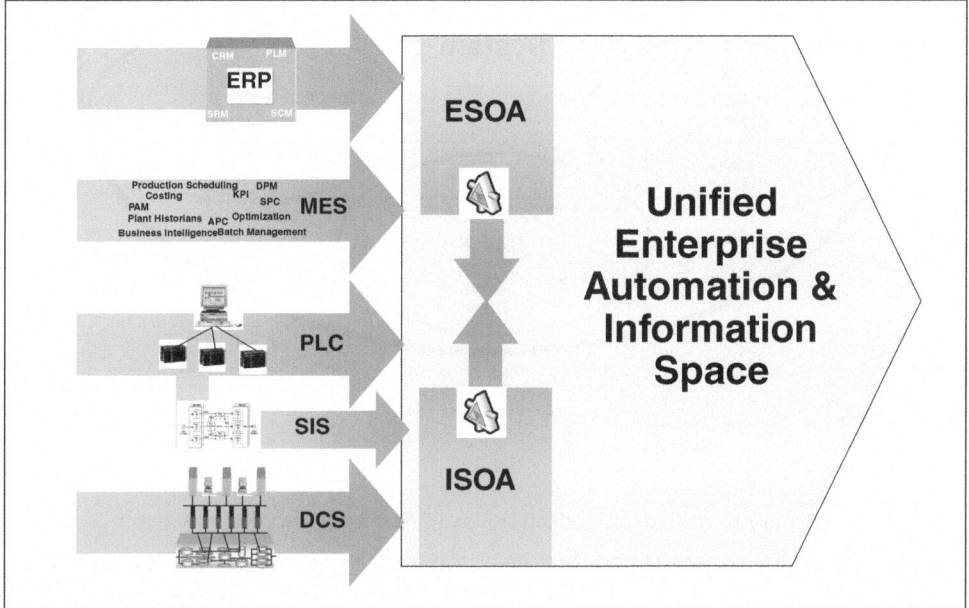

Figure 20-4 Enterprise System Unification

ESOAs and ISOAs that interoperate with each other provide industrial companies with the potential for a problem solution that literally covers entire plants and entire industrial enterprises. In an interesting twist, this enterprise-wide system can be developed using systems and technologies previously installed. The resulting system, comprised of multiple vendor products acquired over many years working as a single system, is what is referred to as an enterprise control system (ECS).

An industrial company with a variety of automation and information systems operating in a number of plants and at corporate can first pull each plant into a single plant-wide ECS, then pull each of the plant systems into the overall business ECS

(Figure 20-5). The result is a system for the entire company that behaves as a single system.

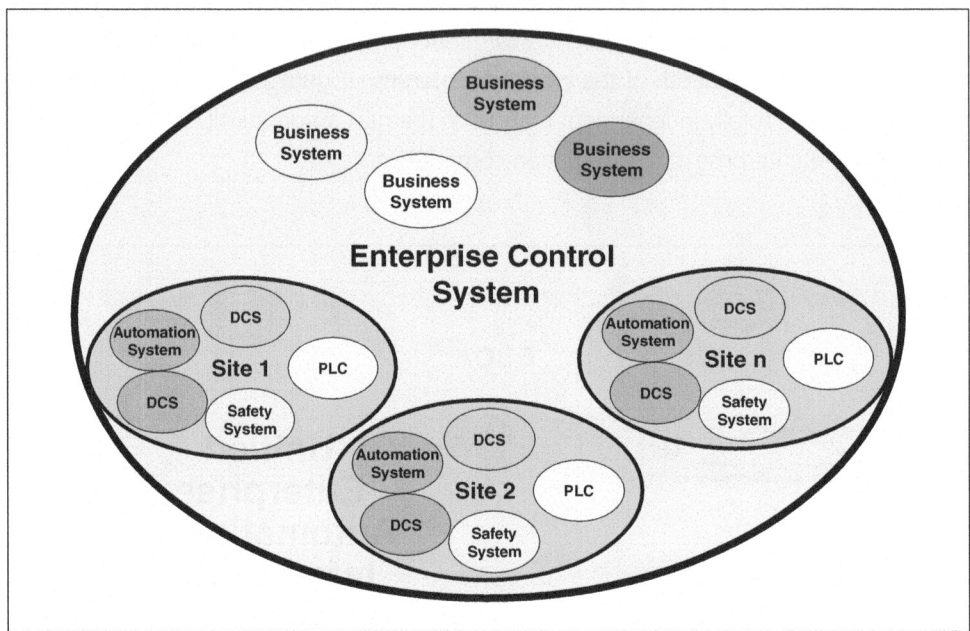

Figure 20-5 Enterprise Control System

As revolutionary to industrial operations as ECSs are, technology by itself seldom provides the expected benefits. As automation systems technology evolved from mechanical systems to pneumatic systems to electronic analog systems to distributed control systems and finally to enterprise control systems, each step along the evolutionary path promised to offer incremental benefits to the users of the technology. And the promise of improvements was real, but seldom realized.

Part of the reason for this was the new systems directly replicated the functionality of the systems they were replacing. Exchanging new technology for older technology that does exactly the same thing seldom provides benefits. Each phase of evolution along the path from mechanical control systems to enterprise control systems offered new functionality and/or new scope. As these systems evolved they provided gains in flexibility, versatility, ease of use, productivity, safety, and cost containment, but the systems were typically installed in a manner so they did not take advantage of any of the potential incremental benefits offered.

This is true for ECSs as much as any of its predecessors. Automation and information technologies are enablers of value, but they do not generate new levels of performance unless the incremental power they offer is utilized. The key to success, therefore, is not merely implementing an ECS. Rather, it is knowing the advantages an ECS offers and then implementing it to accomplish what you can beyond what the traditional systems offered.

The good news is initial experience is starting to show these broad business solutions do provide significant business value. With the business solutions available in ECS environments, perhaps the huge improvements in business performance industrial companies have been hoping for since the dawn of digital computing are finally going to be realized.

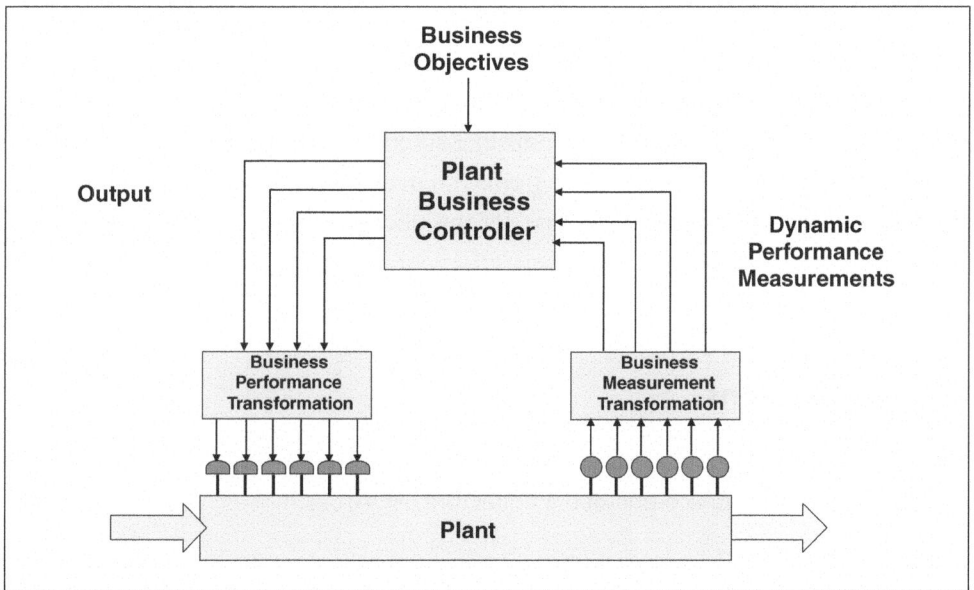

Figure 20-6 Closed Loop Business Control

One trend occurring parallel to the advent of ECS technologies has been the convergence of IT and automation departments. This convergence has caused concern as both departments worry about being taken over by the other. The fact is they have very complementary skills and capabilities that should be able to be used in combination to generate incremental value from industrial operations. For example, automation professionals typically have a good understanding of control theory, while IT organizations may have a better understanding of business issues. By combining these two areas of expertise, the two groups can develop an effective way to apply control theory to business variables.

It is becoming more common for business variables to be measured through the control system in real time. As automation professionals understand, the availability of the measures in the right time frame is the first step to closing the loop on business control (Figure 20-6). The idea of being able to directly and deterministically control business variables in real time offers considerable promise in industrial organizations in which the speed of business is approaching real time. The only way to manage or control a real-time function is with real-time control. Real-time control represents the type of business solution that can result from the implantation of ECS technology and the convergence of intellectual property across the traditional organizational silos.

Advancements and developments are continually taking place in automation and information technology. The evolution of enterprise control systems merely represents one of the latest developments in the industrial marketplace. Truly understanding the importance and impact of these evolutionary steps when they occur requires an understanding of the background of industrial automation and where the various aspects of industrial automation came from. The material previously presented in this book should have provided the basis for this understanding for the emerging enterprise control systems.

Review Questions

1. What is industrial service-oriented architecture (ISOA)?

2. What are some of the unique industrial services that make a service-oriented architecture industrial?

3. Name one example of an enterprise service-oriented architecture (ESOA).

4. What is an enterprise control system (ECS)?

5. Functionally describe closed loop business control.

CHAPTER 21

The Bottom Line: Automation's Business Impact

When it all comes down to it, you can have the greatest team and the greatest technology in the world, but if the team does not apply that technology and push it to the max, you will hurt your bottom line.

Automation technologies applied correctly can have a profound impact on the profitability of an industrial business, but despite the growing potential of advanced automation and information technologies for improving business value, senior management remains reluctant to invest in these technologies. There seems to be a general perception that automation investments seldom realize their expected value. But why are senior managers so skeptical about the value of high technology investments?

From numerous discussions with senior managers, we have discovered business executives believe they have already made huge capital investments in all kinds of high technology but just can't see the business benefits they derived. Executives point out during the approval process for every capital project no one in the operation can confirm the promised returns were actually realized. That is a huge problem.

To analyze the problem, it is helpful to quickly review the basics of the economics of a capital project. Figure 21-1 presents a classic capital economic profile. The bar chart along the bottom of the figure represents the cost associated with the capital investment over the useful life of the investment. The dashed line graph along the top of the figure represents the business benefit realized from the investment. Return on investment (ROI) is simply the integral of the benefit divided by the integral of the cost typically expressed as a percentage over a given time frame. Looking at an automation project after it was completed, in almost every case, the project team could determine the cost of the technology over its lifecycle but had almost no idea of the actual business benefit provided. Cost accounting systems that should provide this information simply don't.

This is because cost accounting systems in industrial operations typically support the financial reporting requirements of the organization and not operations management. Because of this, most cost accounting systems in use today in industrial operations report only monthly, plant-wide cost and margin information. Since so many activities occur in any industrial plant over a month's time, it is impossible to determine which component of any financial improvement or loss should be attributed to any specific activity, even to automation system improvements. Today's cost accounting systems lack the necessary timeliness and detail to be able to effectively measure the benefit of automation systems and technologies. From a business manager's perspective, then, justifying most technology investments requires a leap of faith: They represent a cost with no evidence of discernable benefit.

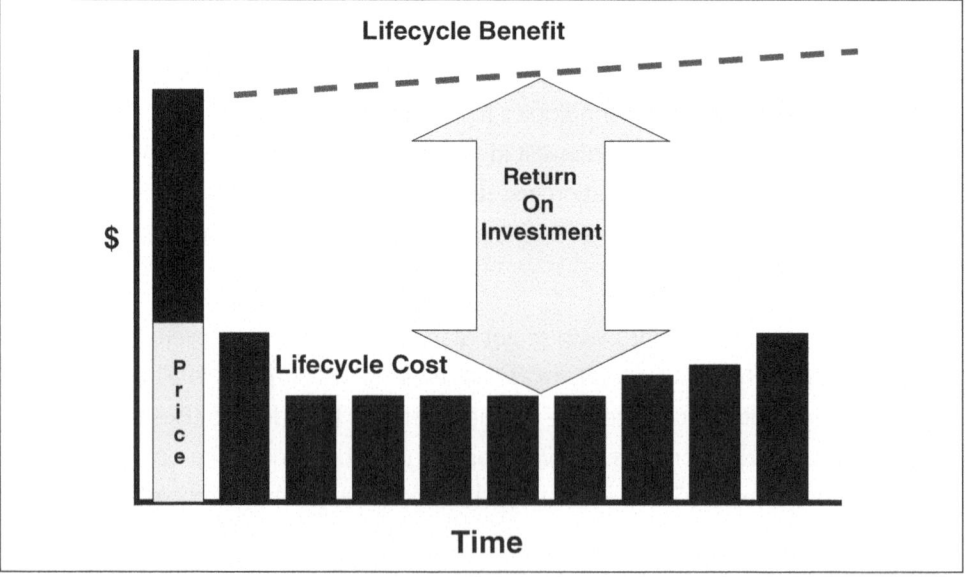

Figure 21-1 Simple Return on Investment

The key word in the previous sentence is "discernable." A few decades ago, the business value of most technology investments seemed to be very clear and measurable. Because automation systems could reduce the number of operators required to run the plant, return on investment was a pretty simple calculation. Over the last decade, however, it has become very apparent there are not enough heads left to reduce to cover the cost of technology. The value of technology investments must now be determined by variables other than personnel reduction. As we have seen, unfortunately, with the current cost accounting systems this has been almost impossible to accomplish.

This inability to measure the business benefit from technology-based capital investments presents a major problem for industrial businesses, but this is not the only technology value problem they face. The second problem is the "replacement technology" mindset, in which new automation systems are acquired to replace aging systems. When this is done, often the new system provides an exact functional replacement of the old system. If, for example, an installed control system begins failing on a regular basis and replacement parts become increasingly expensive and hard to find, the owner typically may issue a request for proposal specifying a replacement system that does exactly what the system being replaced did. Even though the newer system might offer improved and expanded functionality over the system being replaced, the additional functionality would often go unused during the replacement project because it was not initially included as part of the project specification. Project teams are measured by on-time, on-budget delivery of the project as defined in the specification—not on the incremental performance improvement resulting from the project. Although there may be some economic value from the project due to reduced system failures and reduced spare parts costs, the largest opportunity for improved business value is very often missed. Replacing old technology with newer technology that does exactly the same thing seldom delivers much upside business benefit.

It is not that plant managers don't know the new system offers more capability than the one being replaced, they just figure they will take advantage of it once the system is installed and up and running, but this seldom happens. One reason is once the project is commissioned, the technology expertise on the project team that installed the new system leaves to go on to the next project. The talent remaining on site does not always have the capability to utilize the unused functionality of the new system.

Another reason is engineering staffs in industrial plants has been downsized to the point where they are so busy just trying to keep the plant operating they seldom have time to work on the untapped potential of the automation systems. The net result is often less than 40% of the available functionality and capacity of installed automation and information systems is actually utilized during the systems' life. This represents a huge lost opportunity and also contributes to senior management's reluctance to invest in these technologies.

The third problem associated with a lack of perceived business value from automation investments relates closely to the first two. Most industrial companies do not apply continuous improvement-based approaches to automation investments. With all the talk and investment in continuous improvement programs such as Total Quality Management, Six Sigma, and Lean Manufacturing, it would seem natural the concepts associated with programs like this would be applied to getting business value from technology. Instead, we have yet to find a company that is effectively applying their continuous improvement culture to automation technology investments.

One reason for this may be that continuous improvement project teams are primarily measured only by on-time, on-budget delivery, which tends to reward behaviors that are diametrically opposed to continuous improvement. If a system that might provide significant business benefit is not within the initial specification, for example, few project teams would even think about implementing it. It is just not their job, and it may actually cause problems with respect to on-time delivery of the project. But whose responsibility is it? The project team has the expertise, but not the inclination. The plant operations and engineering personnel may have the inclination, but not the expertise. This is a major cultural and performance measurement issue that must be resolved.

Within the context of industry's inability to measure the benefit from technology investments, the replacement technology approach so often employed in industrial operations, and the lack of a continuous improvement culture when it comes to technology investments, it is no wonder senior management has difficulty deciding to invest in automation technology. In today's difficult economic environment, if the business value from technology investments is not both significant and visible, senior management should not be expected to invest.

Changing the Paradigm

Convincing senior management that technology investments are good and beneficial to the business requires a fairly significant paradigm shift. And as we all know, change in the manufacturing industry is not a quick thing. But to begin the shift, you need to find a way to make the benefit from technology investments visible. This necessitates an expansion of the approach to cost accounting commonly employed in industrial operations to include real-time accounting of the benefits as well as the costs. This may seem beyond the authority of most plant personnel, but it is really just an extension of calculations that automation system platforms have performed for decades.

When cost accounting was introduced to manufacturing operations at the dawn of the Industrial Revolution, it utilized a bottom-up approach which accounted for each product as it was produced. As the use of industrial machinery—such as the power loom in the textile industry—increased, production volume increased many-fold, and accounting for the cost of each piece as it was produced became impossible. Instead of accounting for each piece as it was produced, managers began to look at monthly totals, closing the accounting books at the end of each month and measuring results then.

Monthly accounting was a difficult but necessary compromise, which after some time became the normal way of doing business. When computer technology was introduced to manufacturing, finally making it possible to go back and account for the manufacturing operation as it is operating—in real time. Most accountants were too set in their ways to take advantage of it. They had earned college degrees in monthly accounting. It was all they knew. Monthly accounting had become the commonly accepted practice, and that was that. On top of this, most business managers were not aware that such a valuable business function such as real-time accounting was available and they would certainly not look to the automation systems to provide business functionality. And the engineering teams responsible for the automation systems were not typically attuned to accounting requirements.

Fortunately, most major industrial operations are already using automation systems that receive real-time operating data from plant-floor instrumentation. This real-time plant database can be used as the input to real-time accounting models. Engineers with accounting training can model business performance, in real time, right in the automation systems. The same approach can be extended to develop automatic real-time models of all key performance indicators (KPI) of the enterprise. The output of these models can be historized through standard process historians to produce hourly, shift, daily, weekly, and monthly trends in plant accounting. With these in place, the benefit from automation in the capital lifecycle economic model becomes measurable and visible within the operation. This enables the value any improvement activity, such as the implementation of new automation technology, to become visible after the project is completed. Real-time accounting and real-time

KPIs provide the first step in enabling the paradigm shift required to justify technology investments.

The second requirement is industry must move beyond the replacement technology approach commonly employed to this point, to a business value approach. This requires industrial companies to restructure their traditional approach to capital projects, considering the additional value the replacement automation system could provide over the existing systems early enough in the capital project process so the request for proposal (RFP) includes more than just a replacement specification. Industrial companies must also recognize project teams for much more than just keeping to time schedules and budgets; teams must also be recognized for the incremental business value they generate. With the real-time accounting and KPI models in place, it becomes easy to measure the before and after conditions in the plant for every project. The value the project team generates becomes clear and measurable. Senior management must use this to incent project teams to drive improved business value—and the value will come.

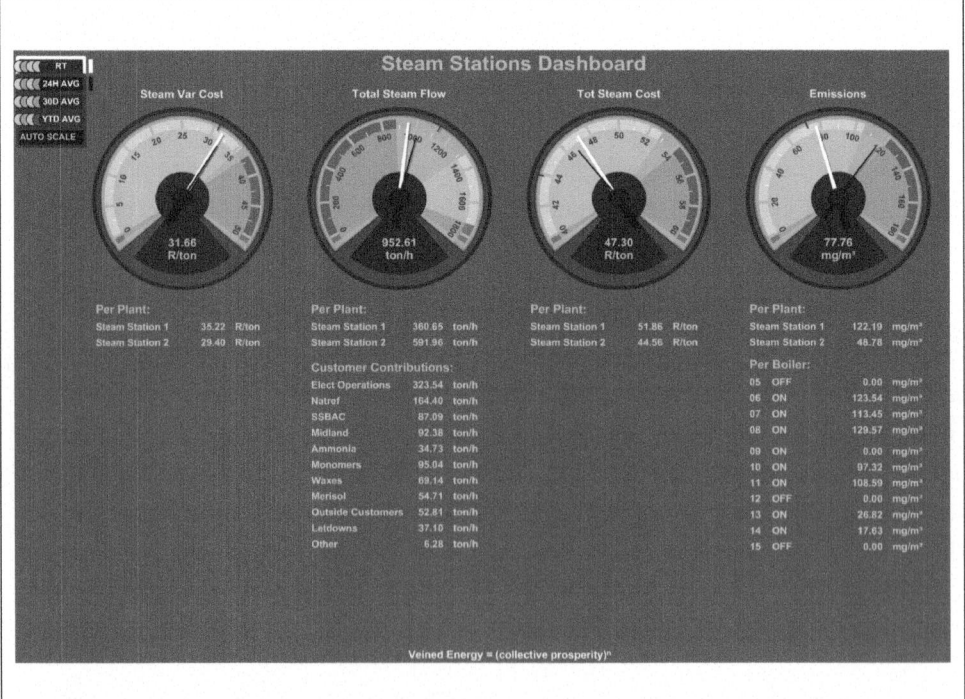

Figure 21-2 Contextualized Performance Dashboard

The improvements must not stop with project completion. Once the project team has moved on, a continuous improvement environment must be in place to maximize and sustain the business value gain. This can be accomplished by using the real-time accounting models and real-time KPIs as the basis for a real-time feedback mechanism for the performance of every person in the operation. Contextualized dashboards (Figure 21-2) for each frontline operator and maintenance person, each engineer, each supervisor, and each manager can be developed to empower every person to make good decisions that continuously drive business value improvements throughout the operation.

When the measures are strategically contextualized to each person's job, they are referred to as dynamic performance measures (DPM). Many industrial operations have tried to develop dashboards for the managers in their plants, which is necessary but not sufficient. It is the frontline operators and maintenance personnel that drive the performance of the operation second-by-second. Industrial companies must start thinking of frontline workers as performance managers who can drive continuous performance improvements.

The combination of real-time performance measures based on real-time accounting and KPI models, a business value-based project approach, and continuous improvement truly defines a totally new paradigm for industrial operations. Employing this approach requires strong leadership. Senior management must not only be part of this change, they must take ownership of it.

Although this culture change and approach are fairly new to industry, where they have been applied the results have been much better than anticipated. The initial results on projects of this type executed in the process industries have realized an average, 100% returns on the capital investments in less than three months. Companies, such as Sasol, Dynegy Midstream Services, and BASF have been publicly recognized in industry publications for the performance of projects executed in this manner. But that is just the beginning, because a continuous improvement culture enables operators, maintenance workers, engineers, and managers to sustain the initial improvements realized through the automation systems and then to drive additional business value improvements from all manufacturing assets. A continuous improvement culture focused on business value encourages cross organizational silo collaboration, which results in even greater value.

It is true senior management has been reluctant to invest in technology, and for good reasons. Traditionally, technology investments have come up way short on the discernable business value they were supposed to generate. If industry continues on the path it has been on, the reluctance to invest in automation technology will most certainly increase. The good news is automation investments can create significant and visible business value improvements. Proving to management that technology investments are worth the cost is critical, but this requires a fundamental shift in the way in which we deal with technology. This change can be enabled by effective performance measures, executing performance-based projects, and implementing a continuous business value improvement culture.

Review Questions

1. What is the primary factor in industrial companies not being able to discern the ROI from automation projects?

2. Why is the actual ROI from automation projects often much less than it could or should have been?

3. Why aren't plant engineers going back and working to utilize the untapped capabilities in automation systems as you might expect them to?

4. How does the availability of dynamic performance measures encourage industrial organizations to take advantage of capabilities in automation systems that are already installed and operating?

Glossary of Terms

Activity-Based Costing (ABC)

An accurate cost management methodology that focuses on indirect costs (overheads) and tracing each expense category to the particular cost object. The basic premise is that cost objects carry out activities that in turn use resources; it is the consumption of these resources that is the driver of cost.

Advanced Process Control (APC)

Process control strategies that go beyond single loop feedback control to provide better control of production processes.

Alarming

Means of drawing the operator or control system's attention to the fact a key process variable has strayed outside acceptable boundaries.

Algorithm

In a control system, a mathematical expression or computational procedure, programmed into a software block, that causes the block to exhibit a predefined characteristic.

Analog Signal

A signal capable of continuous variation over a given range.

Analog to Digital (A to D)

An electronic process in which a continuously variable (analog) signal is changed, without altering its essential content, into a (digital) signal. The input to an analog-to-digital converter (ADC) consists of a voltage that varies among a theoretically infinite number of values. The simplest digital signals have only two states, and are called *binary*. All whole numbers can be represented in binary form as strings of ones and zeros.

Analog Backup

An automation system security approach in which an analog control system would take over control in the event of a failure in the installed digital control system.

Application Program Interface (API)

The specific method prescribed by a computer operating system or by an application program by which a programmer writing an application program can make requests of the operating system or another application.

Asset Availability

The percent of time the asset is available for use. Also, technologies, systems and solutions aimed at maximizing the percent available, minimizing the cost of making the asset available, and preventing catastrophic incidents or failure.

Asset Management

At the industrial plant level asset management typically means the maintenance of the capital assets of the plant. In this usage asset management is equivalent to maintenance management. At the business level in industrial operations, asset management means the business management of any assets in the enterprise asset base.

Asset Utilization

The current output of an asset or asset set divided by the maximum output, usually expressed as a percentage.

Automatic Control

Control that uses automation technology to directly control physical or chemical variables.

Automation

The use of technology to manage and control some of the operations of an industrial plant.

Automation System Security

A class of tools and approaches designed to prevent dangerous or disruptive actions from having an impact on a production process through the automation system.

Availability

> The amount of time a capital asset is operable divided by the total time the capital asset should have been operated over a given time period, usually expressed as a percentage.

Balanced Scorecard (BSC)

> A measurement-based strategic management system, originated by Robert Kaplan and David Norton, which provides a method of aligning business activities to a strategy, and monitoring performance of strategic goals over time.

Batch

> A quantity scheduled to be produced or in production. In non-discrete products, the batch is a quantity planned to be produced in a given time period based on a formula or recipe, which is often developed to produce a given number of end items.

Batch or Discontinuous Process

> A manufacturing process in which fluid-based products are produced in batches or lots.

Batch Management

> The management of a batch process manufacturing operation. Batch management can also refer to the software used to automate some or all of the functions involved in the management of batch process manufacturing operations.

Block Concept

> A software concept designed for process control systems in which a "software block" performs the same function as a piece of analog hardware, enabling engineers with an understanding of analog control to program their digital control system using a similar approach. Software blocks of this type are considered among the earliest of object-oriented software structures.

Break-Fix Maintenance

> See Reactive Maintenance.

Business Intelligence

> Business Intelligence is a process for gathering, processing and disseminating decision-making information to all employees involved with managing the performance of an operation. Business Intelligence is also often used to refer to the information gathered throughout this process.

Cascade Control

A multiple level feedback process control strategy in which a high level controller cascades a set point to a lower level controller.

Capacity

A process dynamic that represents the volume of a physical, chemical or electrical variable that can be stored within the process.

Capital Expenditure

Money spent by a company to add or expand property, plant, or equipment assets, with the expectation that they will benefit the company over a long period of time (more than one year).

Closed Loop Business Control

The application of automatic process control technologies and approaches to business variables.

COBOL

Common Business Oriented Language. A computer programming language designed for business applications.

Computer Integrated Manufacturing (CIM)

Using computers and computer-based systems to coordinate the manufacture of products in a plant, mill or factory.

Computerized Maintenance Management System

Used to handle all aspects of maintenance for key assets, from planning to billing.

Constraint Function

A mathematical model used in a process optimizer that models a constraint to an industrial process.

Continuous Process

Manufacturing process in which raw materials and energy are consumed and products produced in an ongoing and uninterrupted manner once the process is started.

Control System

A hardware/software system that has as its primary function the collection and analysis of feedback from a given set of functions for the purpose of controlling these functions. Control may be implemented by monitoring and/ or systematically modifying parameters or policies used in those functions, or by preparing control reports that initiate useful action with respect to significant deviations and expectations.

Control Theory

In engineering, control theory deals with the behavior of dynamic systems over time. The desired output of a system is called the *reference variable*. When one or more output variables of a system need to show a certain behavior over time, a controller tries to manipulate the inputs of the system to realize this behavior at the output of the system.

Take, for example, cruise control. In this case, the system is a car. The goal of the cruise control is to keep it at a constant speed. So, the output variable of the system is the speed of the car. The primary means to control the speed of the car is the amount of gas being fed into the engine.

A simple way to implement cruise control is to lock the position of the gas pedal the moment the driver engages cruise control. This is fine if the car is driving on perfectly flat terrain. On hilly terrain, the car will accelerate when going downhill and slow down when going uphill, something its driver may find highly undesirable.

This type of controller is called an open-loop controller because there is no direct connection between the output of the system and its input. One of the main disadvantages of this type of controller is its insensitivity to the dynamics of the system under control.

To avoid the problems of the open-loop controller, control theory introduces feedback. The output of the system y is fed back to the reference value r. The controller C then takes the difference between the reference and the output, the error e, to change the inputs u to the system under control P. This is shown in the figure. This kind of controller is a closed-loop controller or feedback controller.

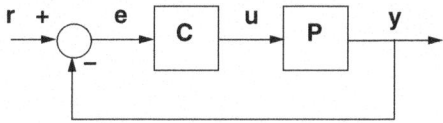

A simple feedback control loop

Customer Relationship Management (CRM)

An information industry term for methodologies, software, and (usually) Internet capabilities that help an enterprise manage customer relationships in an organized way.

Dashboard

A tool used for collecting and reporting information about vital customer requirements and/or your business's performance for key customers. Dashboards are also used to provide a quick summary of process, product and/or business performance.

Data Acquisition System

A system for the automatic collection of data, possibly in real-time, from sensors, instruments and devices in a factory or laboratory, or in the field.

Database

A collection of information. An example is a periodic collection of various process variables that can then be used in reports and trend displays.

Defects Per Unit

The average number of defects observed when sampling a population.
DPU = Total # of defects / Total population

Deterministic

Based on the premise that everything is caused by something, and the outcome of an event or a set of conditions can be predicted because its causes are the same as those of a previous event or set of conditions.

Digital to Analog (D to A)

A process in which signals having a few (usually two) defined levels or states (digital) are converted into signals having a theoretically infinite number of states (analog). A common example is the processing, by a modem, of computer data into audio-frequency tones that can be transmitted over a twisted-pair telephone line. The circuit that performs this function is a digital-to-analog converter.

Digital Signal

A signal that is capable of a limited number of specific values. Discrete values are also referred to as digital values and include both Boolean (two states) and integer values.

Direct Digital Control

A control system in which control devices based on digital computer technology directly send outputs to valves and similar devices without the need to pass through analog control systems.

Discrete Manufacturing Process

A manufacturing process that involves the assembly of parts into products.

Distributed Control System (DCS)

Distributed control systems (DCSs) are used in the process industries controlling breweries, refineries, chemical plants, paper mills, etc. A DCS distributes major control functions, such as controllers, historians and display units (HMI) into different boxes. The key advantage of DCSs is that they divide up the control tasks among multiple distributed systems, so if any single part of a system should fail, the plant could keep operating. DCS also introduced the concept of a data networking, thereby avoiding hard wiring each control point, adding flexibility and reducing the cost of making changes in the production processes.

Dynamic Performance Measures (DPM)

An approach to the development of real-time key performance indicators and real-time accounting measures by modeling these measures from sensor-based data and prioritizing the combined set of real-time measures according to the current strategy of the operation.

Dynamic Simulation

A dynamic simulation consists of a mathematical model of a process plant implemented in a digital computer. This simulation model will include all measurement and control information that would be available to a plant engineer or operator, plus a vast number of additional variables that would not, e.g., the internal hydraulic performance of a distillation column. A dynamic simulation is different from a steady-state simulation in that a dynamic simulation predicts how process variables change with time when moving from one steady state to another, or during a transient upset; a steady-state simulation only shows the values of the variables when the plant is in a steady state.

Efficiency

A term indicating the optimization of productivity (measured outputs over measured inputs) typically stated on a 0–100% scale. To improve efficiency, the productivity ratio must be improved (the input to output ratio must be decreased). See productivity.

Electromechanical Relay

An electrical device that performs as an automatic switch through the use of an electromagnet and other electrical components.

Electronic Analog Control System

An automatic analog control system powered by electricity and using electronic components such as PLCs.

Enterprise Asset Management (EAM)

Management of the entire lifecycle of assets, from acquisition to retirement. EAM involves the effective management of multiple resources for maintenance, repair & operations (MRO) including detailed instructions, labor/skill requirements and parts.

Enterprise Control System (ECS)

A control approach that enables manufacturers to develop solutions that span their business enterprises without concern for the constraints traditionally imposed by crossing the boundaries of the different classes of systems. The following four characteristics define an ECS: full plant-floor interoperation; open communication access across the business enterprise; support for asset performance management (APM) tools which enable unified maintenance and operations management; and a unified engineering environment across all plant-floor domains.

Enterprise Information System

These systems provide the information infrastructure for an industrial enterprise. Enterprises run their businesses using the information stored in these systems; the success of an enterprise critically depends on this information.

Enterprise Resource Planning (ERP)

An industry term for the broad set of activities supported by multi-module application software that helps a manufacturer or other business manage the important parts of its business, including product planning, parts purchasing, maintaining inventories, interacting with suppliers, providing customer service, and tracking orders. ERP can also include application modules for the finance and human resources aspects of a business.

Exception Logic

Software in a batch management system that determines the appropriate action for the automation system to take upon recognition of an unexpected event during the manufacture of a batch.

Factory Automation

Computer-based automation for the main activities of factory production, including procurement, control, production planning and communications.

Fault Tolerance

The ability to identify and compensate for failed control system elements and allow repair while continuing an assigned task without process interruption. Fault tolerance is achieved by incorporating redundancy and fault masking.

Feedback Control

A control strategy in which the controller reacts to correct the process upon a deviation in a measured variable.

Feedback Trim

A feedback adjustment to a feedforward process control strategy.

Feedforward Control

A control strategy in which the controller predicts that a process deviation will occur if corrective action is not taken and executes the corrective action at the right time to prevent the deviation.

Feedstock

The primary raw material in a chemical or refining process, normally received by pipeline or in large-scale bulk shipments.

Firewall

A system security tool that is designed to prevent potentially disruptive messages and communications from entering the system.

FORTRAN

Formular Translator—a computer programming language developed for scientific applications.

Historian

A real-time database used to capture information about process plants over extended periods of time. Historian applications can also be used to analyze the collected data to identify problems and trends in order to improve plant performance.

Human-Machine Interface (HMI)

A general term in human-computer interaction for the interface between a user and a computer, of which a graphical user interface (GUI) is a part. Also, a software tool for visualizing and controlling automation processes. Also, a general term referring to any interface between a human and a machine or piece of equipment.

Industrial Automation

A system or set of systems for automating manufacturing processes and operations.

Input/Output (I/O)

I/O describes any operation, program, or device that transfers data to or from a computer. Typical I/O devices are printers, hard disks, keyboards, and mice. In fact, some devices are basically input-only devices (keyboards and mice); others are primarily output-only devices (printers); and others provide both input and output of data (hard disks, diskettes, writable CD-ROMs). In industrial operations input/output may also refer to the signal coming from and going to the manufacturing process.

ISA-95

A standards approach championed by ISA that defined the interoperation between industrial plants and enterprises.

Key Performance Indicator (KPI)

Key performance indicators are predefined measures that provide high-level snapshots of an operation. KPIs typically consist of any combination of reports, spreadsheets, or charts. They may include global or regional sales figures and trends over time, personnel stats and trends, real-time supply chain information, or anything else that is deemed critical to a corporation's success.

Laboratory Information Management System (LIMS)

A computer-based system designed to automate many of the activities associated with laboratory quality analysis in industrial operations.

Ladder Logic

Ladder logic provides a graphical representation of a logic control strategy, composed of contacts, coils, counters and timers, that is rapidly processed in programmable logic controllers.

Loop Tuning

The process of setting the proper adjustments (proportional, integral and derivative) in a feedback controller to realize the desired control loop response from the controller.

Maintenance

The work and activity associated with the maintenance of the company's capital assets.

Manual Backup

An automation system security approach in which an operator can switch to the manual operation of a process at his or her discretion.

Manual Control

A mode of control in which a human operator controls the process.

Manufacturing

The making or processing of raw material into finished products, especially by large scale industrial processes.

Manufacturing Execution Systems (MES)

Software applications of industrial operations that operate between DCS and ERP systems. Applications include operations monitoring/logbook, target setting, scheduling, laboratory systems, asset management, quality monitoring and control, batch/lot tracking, APC, optimization, recipe management, blending control/monitoring/optimization, safety monitoring, equipment monitoring and others.

Manufacturing Process

A set of operations designed to convert incoming materials into outgoing products through the utilization of basic production resources such as equipment, tools, energy and manpower.

Manufacturing Resource Planning (MRP)

MRP is a method for the effective planning of all resources of a manufacturing company. MRP systems typically include business planning, sales and operations planning, production scheduling, material requirements planning, capacity requirements planning, and the execution support systems for capacity and material. Output from these systems is integrated with financial reports such as the business plan, purchase commitment report, shipping budget, and inventory projections in dollars.

Master Recipe File

A computer file with the general recipes for all major products produced through batch manufacturing operations across an industrial enterprise.

Master Terminal Unit

The master computer system in a SCADA system.

Mechanical Control Systems

Process control systems built on mechanical mechanisms.

Mean Time Between Failure (MTBF)

The expected time between failures of a system or piece of equipment, based on historical records or vendor statistics.

Mean Time To Repair (MTTR)

The expected time to repair a failed system or piece of equipment—usually expressed in hours.

Measurement

The present value of a process variable (e.g., flow rate, pressure, temperature or liquid level).

Multivariable Predictive Control

A control strategy that measures multiple process variables and uses a dynamic process model to drive the process in an optimum direction by manipulating several variables simultaneously.

Normally Closed Relay

An electromechanical relay that is in the closed position when power is not being applied to the coil.

Normally Open Relay

An electromechanical relay that is in the open position when power is not being applied to the coil.

Object Oriented

The use of a class of programming languages and techniques based on the concept of an "object," which is a data structure (abstract data type) encapsulated with a set of routines, called "methods," which operate on the data. Operations on the data can only be performed via these methods, which are common to all objects that are instances of a particular "class." Thus the interface to objects is well defined, and allows the code implementing the methods to be changed so long as the interface remains the same. Moving objects between different applications enables the reuse of code.

Objective Function

A mathematical model that defines the objective of an optimization problem.

Open Control System

A control system that is able to work with components from many different vendors. Typically, an open control system will provide:

- A wide array of popular interfaces
- A compatible environment for commercial off-the-shelf software
- Multi-vendor control architectures
- Rapid adaptation of business protocols and communications standards.

Operate To Breakdown

A maintenance strategy which operates equipment until it breaks down.

Operating System

An operating system (sometimes abbreviated as "OS") is the program that, after being initially loaded into a computer by a boot program, manages all the other programs in the computer.

Operations Business Excellence

A strategy for industrial operations involving the convergence of operations excellence, business excellence and strategic excellence into a single set of metrics and improvement initiatives.

Operations Equipment Excellence (OEE)

A composite performance measure of operations excellence that is the product of asset availability, product rate and quality over a given period of time, typically expressed as a percentage. OEE is often the primary measure used to determine continuous improvement for an operations excellence strategy.

Operations Excellence

A traditional operational strategy involving continuous improvement of industrial operations through the management of plant operations and asset management, and typically measured by key performance indicators (KPI), often with a single overall composite KPI (typically operations equipment excellence [OEE]) for benchmarking and reporting.

Operator Training Simulator (OTS)

A software simulation system for an industrial operation that enables the training of operators in operating environments that appears to be very close to reality.

Optimization

Sophisticated software designed to either minimize or maximize a single objective associated with a process based on a set of constraints.

Overall Equipment Effectiveness (OEE)

A composite key performance indicator that measures the impact the current performance of any individual piece of equipment, such as throughput or downtime, has on the overall efficiency of the plant. OEE = Availability x Performance Efficiency (production rate) x Yield (quality)

Phase of Operation

A major processing step in a batch manufacturing process.

Phase Logic

Software that defines and executes the operation of a batch through a set of equipment.

Piping and Instrumentation Diagram (P&ID)

A schematic illustration of the functional relationship of piping, instrumentation and system equipment components. A P&ID shows all of piping, including the physical sequence of branches, reducers, valves, equipment, instrumentation and control interlocks. The plant's P&IDs are used to operate the process system.

Planning and Scheduling Software

Software that determines which products to make and on what time schedule in industrial operations.

Plant Asset Management (PAM)

PAM is the overall coordination of the various maintenance strategies used in industrial plants, including break-fix, preventive maintenance, predictive maintenance and RCM, utilized in industrial plants as well as maintenance scheduling and spare parts management.

Plant Information Management System

The Plant Information Management System is automation software that delves into the uses of real-time and near real-time information in current business applications and operational decision-making processes.

Positioner

A device that ensures that the closing or throttling element of a valve moves to or maintains the correct position.

Pneumatic Control System

An automatic analog process control system designed to operate on air power.

Predictive Maintenance

The detection of the signs of early degradation in a device to repair a problem before it actually occurs.

Preventive Maintenance

Carrying out of time-based regular maintenance in order to prevent breakdown before it occurs.

Process

A set of chemical or physical tasks or combinations of tasks, performed in serial or parallel, to produce fuels, chemicals, food, pharmaceuticals, paper, electrical power, or other finished products.

Process Automation System

A system to control complex production processes in various industries.

Process Control Equipment

Equipment that measures the variables of a technical process, directs the process according to control signals from the process computer system, and provides appropriate signal transformation. Examples of process control equipment include actuators, sensors, and transducers.

Process Control Loop

A feedback loop for the control of a continuous process comprised of the process, an instrument to measure the controlled variable, a controller, and a valve or devices such as dampers that control air flow and motor drives that control the speed of motors.

Process Control System

A system consisting of a controller, process control equipment and a process interface that will maintain a process within certain parameters of key variables, taking action to rectify deviations.

Process Manufacturing

Production that adds value by mixing, separating, forming, and/or performing chemical reactions. It may be done in either batch or continuous mode.

Process Safety Management

Process safety management is the effective way to deal with operations that store, handle, or process toxic or flammable materials in quantities that, if released, could have a major impact on workers, nearby communities, or facilities. These events can have significant life, safety, environmental, legal, regulatory, and financial consequences.

Process Simulation Optimization

Improving plant performance by means of simulation.

Process Variable

The level, quantity, other condition in the process that is to be directly measured and controlled.

Productivity

The ratio of measured outputs over measured inputs (i.e., widgets produced per man-hour).

Programmable Logic Controller (PLC)

A computer-based logic controller designed to replace relay ladder logic systems for the control of machinery in industrial operations. PLCs utilize a ladder logic programming language designed to replicate the behavior of relay ladder logic systems.

Purdue CIM Reference Model

> A logical model of the functions of automation and business systems developed by a team of professionals led by Dr. Theodore Williams of Purdue University. Although this is an extensive model it is commonly represented in a simple five layer functional model showing a hierarchy of automation and information functions of industrial businesses.

Recipe

> A procedure and set of values (formula) that guide the making of products in batches.

Relay Ladder

> A collection of interconnected electromechanical relays designed to work together to perform one or more of a set of logic control functions.

Reliability Centered Maintenance (RCM)

> A strategy for the implementation of equipment and maintenance approaches that involves the design, reliability and failure impact of each major piece of equipment in an operation to determine the probability of a breakdown in the equipment over time and the business impact to the operation if a breakdown were to occur.

Quality Assurance (QA) includes the following meanings:

- All actions taken to ensure that standards and procedures are adhered to and that delivered products or services meet performance requirements.

- The planned, systematic activities necessary to ensure that a component, module, or system conforms to established technical requirements.

- The policy, procedures, and systematic actions established in an enterprise for the purpose of providing and maintaining a specified degree of confidence in data integrity and accuracy throughout the lifecycle of the data, which includes input, update, manipulation, and output.

Quality Control (QC)

In manufacturing, quality is sometimes defined as "meeting the requirements of the customer." The term quality control or sometimes quality assurance describes any systematic process for ensuring quality during the successive steps in producing a product.

Reactive Maintenance

The "run it till it breaks" maintenance mode. No actions or efforts are taken to maintain the equipment as the designer originally intended, either to prevent failure or to ensure that the design life of the equipment is reached.

Real Time

Real time is a level of computer responsiveness that is completed in a time frame relative to an external process (for example, to present visualizations of the weather as it constantly changes) with which the computer is interacting. Real-time is an adjective pertaining to computers or processes that operate in real time. Real time is also associated with the management of business processes at the same rate as the related manufacturing processes are taking place.

Real-time Business Intelligence

Business intelligence made available to decision makers in real time.

Real-time Process Management

Mechanisms to enable managers not only to go beyond simply monitoring key data in real time, but also to understand the process implications and be able to respond proactively, and in some cases automatically.

Reliability Centered Maintenance (RCM)

A structured maintenance decision-making tool that allows primarily for planned maintenance activities. The result is a shift in maintenance resources to areas of greatest effect in ensuring maintenance productivity.

Remote Terminal Unit (RTU)

In SCADA systems, an RTU is a device installed at a remote location that collects data, codes the data into a format that is transmittable and transmits the data back to a central station, or master terminal unit. An RTU also collects information from the master device and implements processes that are directed by the master. RTUs are equipped with input channels for sensing or metering, output channels for control, indication or alarms and a communications port.

Safety Instrumented System (SIS)

A computer-based system that identifies impending unsafe events and responds in a manner to protect life, equipment and product.

Safety Integrity Level (SIL)

The SIL is a measure of system criticality and defines the safety performance criteria for the system.

SIL	Qualitative View of SIL
4	Catastrophic Community Impact
3	Employee and Community Impact
2	Major Property and Production Protection. Possible Injury to Employee
1	Minor Property and Production Protection

The assignment of SIL is a company decision based on risk management and risk tolerance philosophy. Best Practices encourage companies to have consistent SIL applications between similar process units throughout a company's fleet of plants.

Safety Shutdown System

An automation system designed to monitor a dangerous manufacturing process or plant and to shut the plant down in a safe manner upon detection of an impending event.

Service-Oriented Architecture (SOA)

SOA refers to a portfolio of loosely-coupled, network addressable business services. These Services are programs that 1) communicate by exchanging well-understood messages and 2) are composed of a set of components which can be invoked and whose interface descriptions can be published and discovered.

Set Point

An input variable that establishes the desired value of the process variable being controlled.

Set Point Control System

A digital computer system over an electronic analog control system, in which the digital system sets the appropriate set points of the analog controllers.

Set Point Tracking

An automation system security approach in which the set points in a backup analog control system are periodically set to the value of the corresponding set points in the controlling digital system; thus, upon a failure in the digital system, the backup analog system will take over control in a smooth manner.

Simulation

Computer-based software or other mechanism that behaves in a manner very similar to a physical process or set of processes.

Single Loop Controller

A controller that controls a very small or a critical process. A single loop controller measures technological parameters such as temperature, pressure, level, and flow. These controllers accept direct signals from thermoresistors or thermocouples, as well as unified electric current and voltage signals. The measured variables are displayed in their respective units.

Six Sigma

Six Sigma is a disciplined, data-driven approach and methodology for eliminating defects (driving toward six standard deviations between the mean and the nearest specification limit) in any process—from manufacturing to transactional and from product to service. The statistical representation of Six Sigma describes quantitatively how a process is performing. To achieve Six Sigma, a process must not produce more than 3.4 defects per million opportunities. A defect is defined as anything outside of customer specifications. A Six Sigma opportunity is then the total quantity of the chances for a defect.

Statistical Process Control (SPC)

A variety of statistical techniques for measuring, analyzing, improving and controlling processes and for ensuring that they operate with minimum variance from standard operating conditions.

Statistical Quality Control (SQC)

The application of statistical techniques to control quality.

Supervisory Control And Data Acquisition (SCADA)

A computer system for gathering, analyzing and responding to real-time data over great distances. SCADA systems are used to monitor and control a plant or equipment in industries such as telecommunications, water and waste control, energy, oil and gas refining and transportation.

Supply Chain Management (SCM)

The oversight of materials, information, and finances as they move in a process from supplier to manufacturer to wholesaler to retailer to consumer. Supply chain management involves coordinating and integrating these flows both within and among companies. It is said that the ultimate goal of any effective supply chain management system is to reduce inventory (with the assumption that products are available when needed).

Systems Integration

Combining sub-systems and/or peripherals, adding software and cabling to specification in order to produce fully configured system.

Systems Integrator (SI)

An individual or company that specializes in systems integration—the building of complete computer systems by putting together components from different vendors.

Theory of Constraints (TOC)

Also called constraints management, a management philosophy developed by Dr. Eliyahu M. Goldratt that is broken down into three interrelated areas: logistics, performance measurement, and logical thinking.

Total Quality Management (TQM)

TQM is a systematic, organization-wide approach that motivates, supports, and enables quality management in all activities, focusing on the needs and expectations of internal and external stakeholders.

TÜV

A German company headquartered in München, offering various services around quality and accreditation in different industries. It operates worldwide.

Variance Report

Financial analysis report traditionally produced on a monthly basis by the financial system to overview the financial performance of manufacturing operations. The primary financial statistic presented on these reports is cost/(units of product made) as measured against the expected cost per unit.

Vollmann Strategy Decomposition

A structured process developed by Dr. Thomas Vollmann for the decomposition of production strategy through an industrial operation.

Work in Process (WIP)

A product or products in various stages of completion throughout the plant, including all material from raw material that has been released for initial processing up to completely processed material awaiting final inspection and acceptance as finished product. Many accounting systems also include the value of semi-finished stock and components in this category. Synonym: In-process inventory.

Bibliography

Abel, J. and Martin, P. "Automation, Business, and Operating Advances Align into New Paradigm for Economic Performance Improvement." *Pharmaceutical Engineering*, Tampa: ISPE, Jan./Feb. 2000.

Allen, L. and Stovicek, D. "CIM Goes to School." *Automation*, Cleveland: Penton Publishing, May 1990.

Anderson, N.A. *Instrumentation for Process Measurement and Control*. Radnor: Chilton Company, 1980.

Andrews, W. G. and Williams, H. B. *Applied Instrumentation in the Process Industries*. Houston: Gulf Publishing Company, 1979.

Babb, M. "Process Control Systems in the 1990s Will Enter a 'New Age.'" *Control Engineering*, Newton: Cahners Publishing, Jan. 1990.

Badavas, Dr. P. C. and Epperly, Dr. A. D. *Statistical Process Control Integrated with Distributed Control Systems*. NPRA Computer Conference, Pittsburgh, Oct. 1988.

Baer, T. "The Software Side of Quality." *Managing Automation*, New York: Thomas Publishing Company, May 1990.

Balch, B. and Miller, T. "Statistical Process Control in Food Processing." *Intech*, ISA, Jul. 1990.

Beaverstock, M. and Byzyna, L. "Measuring Performance Dynamically." *Control Engineering*, Newton: Cahners Publishing, Vol. II, Jun. 1990.

Benassi, F. "The Long Road to Quality." *Managing Automation*. New York: Thomas Publishing Company, May 1990.

Box, G. E. P.; Hunter, W. G.; and Hunter, J. S. *Statistics for Experimenters: An Introduction to Design, Data Analysis, and Model Building*. New York: John Wiley & Sons, 1978.

Brimson, J. A. "How Advanced Manufacturing Technologies Are Reshaping Cost Management." *Management Accounting*, Mar. 1986.

Brown, M. E.; Parker, K.; and Senyard, C. P. Jr. "Computer Integrated Manufacturing in the Chemical Process Industries." *Chemical Processing*, Oct. 1989.

Bunnell, J. and McAllister, R. L. "Continuous Improvement Through Plantwide Integration." *Control Engineering*, Newton: Cahners Publishing, Volume II, Jun. 1990.

Bylinsky, G. "Challengers Move In On ERP." *Fortune*, New York: Time Life Inc., Nov. 22, 1999.

Cargill, C. F. *Information Technology Standardization*. Maynard: Digital Press, Digital Equipment Corporation, 1989.

Clark, K. B. and Hayes, R. H. "Why Some Factories Are More Productive Than Others." *Harvard Business Review*, Harvard Business Press, Cambridge: Sep.-Oct. 1986.

Conneally, D., Lovett, D., and Martin, P. G. "An Econometric Approach to Justifying Advanced Process Control Projects." *Control Solutions*, PennWell, Mar. 2001.

Cooper, N. and Martin, P. G. "The Road to Asset Performance Management." *Plant Services Webcast*, Feb. 2007.

Cooper, R. and Kaplan, R. S. "Measure Costs Right: Make the Right Decisions." *Harvard Business Review*, Harvard Business Press, Cambridge: Sep.-Oct. 1988.

Cooper, R. and Martin, P. G. "Real-Time Cost Accounting: From the Factory Floor to the Bottom Line." *Baan Webcast*, Baan, Oct. 2001.

Cox, B. J. *Object Oriented Programming*, An Evolutionary Approach. Reading: Addison-Wesley Publishing Company, 1986.

Crosby, P. B. *Quality is Free: The Art of Making Quality Certain*. New York: McGraw-Hill Book Company, 1979.

Deming, W. E. *Out of the Crisis*. Cambridge: MIT Center for Advanced Engineering Study, 1986.

Dixon, J. R.; Nanni, A. J.; and Vollmann, T. E. *The New Performance Challenge: Measuring Operations for World Class Competition: (Irwin/Apics Series in Production Management)*. McGraw-Hill Professional Publishing, 1990.

Drucker, P. F. "The Emerging Theory of Manufacturing." *Harvard Business Review*, Harvard Business Press, May-Jun. 1990.

——— *Innovation and Entrepreneurship: Practices and Principles*. New York: Harper and Row, 1985.

——— *The Practice of Management*, New York: Harper & Row, 1954.

Elwart, S. P. and Martin, P. G. "New Software Structures Extend Control Capabilities." *Control Engineering*, Newton: Cahners Publishing, June 1990, Volume II.

Foxboro Company, The. *Introduction to Process Control*, Foxboro: Foxboro Company, 1986.

Friedman, P. G. *Economics of Control Improvement*. Research Triangle Park: ISA, 1995.

Gagne, J. *Quality Performance Means More at Dow*. Midland: Dow Chemical U.S.A.

Ghosh, A. and Rosenof, H. P. *Batch Process Automation: Theory and Practice*. New York: Van Nostrand Reinhold Company, Inc., 1987.

Goldratt, E.M. *What is This Thing Called Theory of Constraints and How Should It Be Implemented?* Great Barrington: North River Press, 1990.

Gould, L. "The CIM Task Force: Purchasing CIM is Also a Distributed, Hierarchical Pursuit." *Managing Automation*, Thomas Publishing, Dec. 1990.

Grant, E. L. and Leavenworth, R. S. *Statistical Quality Control, Fifth Edition*. New York: McGraw-Hill Book Company, 1980.

Harbor Research "Performance Measurement: Impact on Competitive Performance." *Outlook*, Boston: Harbor Research Corp., Vol. 6, No. 4 (1990).

Harrold, D. "Enterprise Integration Requires Understanding of the Plant Floor." *Control Engineering*, Newton: Cahners Publishing, Feb. 2000.

Harry, M. J., Ph.D. *The Vision of Six Sigma: A Roadmap for Breakthrough*. Tempe: Six Sigma Academy, Inc., 1994.

Hays, W. L. *Statistics for the Social Sciences*. New York: Holt, Rinehart and Winston, Inc., 1973.

Henderson, B. A. and Larco, J. L. *Lean Transformation*. Richmond: The Oaklea Press, 1999.

Hidden, A. E. and Lowe, E. *Computer Control in the Process Industries*. London: Peter Peregrinus Ltd., 1971.

Jasany, L. C. "Knowledge (and Power) to the People." *Automation*, Cleveland: Penton Publishing, Jul. 1990.

Jones, D. T.; Roos, D.; and Womack, J. P. *The Machine That Changed the World: The Story of Lean Production*. New York: Harper Perennial, 1991.

Juran, Dr. J. *Juran on Leadership for Quality - An Executive Handbook*. New York: The Free Press, 1989.

———— *Managerial Breakthrough: A New Concept of the Manager's Job*. New York: McGraw-Hill Book Company, 1964.

———— "One Cost System Isn't Enough." *Harvard Business Review*, Harvard Business Press, Cambridge: Jan.-Feb. 1988.

———— "Yesterday's Accounting Undermines Production." *Harvard Business Review*, Harvard Business Press, Cambridge: Jul.-Aug. 1984.

Kompass, E. J. "The Road to Plantwide Information." *Control Engineering*. Newton: Cahners Publishing, Volume II, Jun. 1990.

Lareau, A. F. "Bringing Information to Automation." *Control Engineering*, Newton: Cahners Publishing, Volume II, Jun. 1990.

MacGregor, John F. "On-Line Statistical Process Control." *Chemical Engineering Progress*, Putman Media, Oct. 1988.

Martin, P. G. "A Leadership Dilemma." *InTech*, ISA, Feb. 2008.

———— "A New Blueprint for Automation." *Chemical Plants and Processing*. Sep. 1989.

———— "A Software Structure for Open Industrial Systems." *Proceedings of the Fourteenth Annual Advanced Control Conference*, Purdue University and Control Engineering, Sep. 1989.

———— "Achieving Process Manufacturing Excellence." *IndustryWeek Webcast*, Nov. 2007.

———— "Automation Outlook." *Start*, Feb.2002.

———— "Automation Payback." *Industrial Computing*, Research Triangle Park: ISA, Aug. 1999.

———— "Bottom-Line Automation." Conference Paper, ISA, Oct. 2001.

Martin, P. G. and Turk, M. "Bottom to Top Real Time Accounting and Performance Measures: From Plant Floor to ERP." *NPRA*, Oct. 2008.

Martin, P. G. "Computer Control of Batch Processes." *Measurement and Control*, Issue 106, Sep. 1984.

———— "Designing a Batch Control System." *I&CS: Control Technology for Engineers and Engineering Management*, Radnor: Chilton Company, Oct. 1987.

———— "Don't Fumble Your Tech Funding Opportunity." *Pharmaceutical Manufacturing*, Putman Media, Jan. 2009.

———— "Driving Bottom Line Value Through Enterprise Control." *Heartbeat*, Invensys, Dec. 2008.

———— "Dynamic Performance Management: The Path to World Class Manufacturing." New York: Van Nostrand Reinhold, 1993.

———— "Interoperability is Not the Solution." *InTech*, Oct. 2005.

———— "Living the Dream of Plant Asset Management." *InTech*, Oct. 2005.

———— "Manufacturing Intelligence through Enterprise Control." *Manufacturing Business Technology Webcast*, Oct. 2007.

———— "Metrics That Matter." *Managing Automation Webcast*, Jun. 2005.

———— "Offene Industriesysteme fur die Verfarhrenstechnik-Schritt in ein neues Automatisierungszeitalter." *Verfahrenstechnik*, Nr. 10 (1989).

———— "OIS: A Blueprint for Automation." *Control Engineering*, Newton: Cahners Publishing, Feb. 1989.

———— "Open Communications in the Process Industries." Paper presented at ISA, Houston Channel Section, Apr. 1990.

———— "Open Industrial Systems." *Measurement and Control*, Oct. 1989.

———— "Out of Many … One: Pointing the Whole Organization in the Same Direction." *UpTime*, Jan. 2009.

———— "Real-Time Performance Measurement and Management: From the Plant Floor to the Bottom Line." *Conference Proceedings, AIChE 2003 Spring National Meeting*, Apr. 2003.

———— "Temperature, Pressure, Flow, Profit." *InTech*, Feb. 2001.

———— "The Benefits and Value of Automation." *ISA*, Oct. 2008.

———— "The Business of Engineering." *Industrial Management*, Jan./Feb. 2004.

———— "The Information Empowered Enterprise: Driving Operational Excellence." *Control Engineering*, Jan. 2008.

———— "The Move to Open Communications in the Process Industries." Paper presented at ISA, Calgary Section, Apr. 1990.

———— "The Move Toward Open Industrial Systems." *Proceedings of the Advanced Control Symposium*, College Station: A&M University, Feb. 1989.

———— "The Move to Performance-based Automation." *InTech*, Sep. 2004.

———— "Time for Leaders to Step Forward." *InTech*, Sep. 2008.

Martin, P. G. and Bapat, V. "The Pursuit of the Perfect Plant: Enabling Real Time Production Responsiveness Through Enterprise Control." *Automation World Webcast*, Oct. 2007.

Maskell, B. H. "Performance Measurement for World Class Manufacturing, Part I." *Manufacturing Systems*, Carol Stream: Hitchcock Publishing Co., Jul. 1989.

————— "Performance Measurement for World Class Manufacturing, Part II." *Manufacturing Systems*, Carol Stream: Hitchcock Publishing Co., Aug. 1989.

McLuhan, M. *The Mechanical Bride: Folklore of Industrial Man*. Boston: Beacon Press, 1952.

McSween, T. and Zaloom, V. Jr. "Creating a Positive Work Environment." *Chemical Engineering*, Jun. 1990.

Miller, J. G., Nianni, A. J., and Vollmann, T. E. "What Shall We Account For?" *Management Accounting*, Jan. 1988.

Miller, J. G. and Vollmann, T. E. "The Hidden Factory." *Harvard Business Review*, Harvard Business Press, Cambridge: Sep.-Oct. 1985.

Mitchell, J. S. *Physical Asset Management Handbook, Fourth Edition*. Houston: Clarion Technical Publishers, 2006.

Murrill, P. W. *Fundamentals of Process Control Theory*. Research Triangle Park: ISA, 1981.

Passino, R. "Computers Spur Profitability at General Chemical." *Chemical Processing*, Chicago: Putman Publishing, Nov. 1990.

Pastor, E. "Networks: The Backbone of Integration." *Automation*, Cleveland: Penton Publishing, May 1990.

Peat, F. D. *Artificial Intelligence: How Machines Think*. New York: Simon & Schuster, 1985.

Penton Publishing. "A Special Report on Factory Automation: Automation's Global Game." *Industryweek*, Cleveland: Penton, Jun. 20, 1988.

Penton Publishing. "A Special Report on Factory Automation: Linking the Pieces of CIM." *Industryweek*, Cleveland: Penton , Mar. 23, 1987.

Penton Publishing. "A Special Report on Factory Automation: The U.S. and Quality: A New Culture." *IndustryWeek*, Cleveland: Penton, Apr. 17, 1989.

Penton Publishing. "A State-of-the-Art Report: Factories of the Future." *IndustryWeek*, Cleveland: Penton, Mar. 21, 1988.

Perrow, C. "Normal Accidents: Living with High-Risk Technologies." Princeton: Princeton University Press, 1999.

Peters, T. J. and Waterman, R. H., Jr. *In Search of Excellence: Lessons from America's Best-Run Companies*. New York: Harper & Row, 1979.

Quak, A. "Batch Control Systems." *State-of-the-Art Newsletter*, Foxboro: The Foxboro Company, May 1986.

Qualtec, Inc. *Managing Quality Improvement, Participant Workbook*. Miami: Florida Power & Light Co., 1988.

Qualtec, Inc. *Team Leader Training Course, Participant Workbook*. Miami: Florida Power & Light Co., 1987.

Rohan, T. M. "Factories of the Future." *IndustryWeek*, Penton Media, Mar. 21, 1988.

Rossi, D. A. "Consider Integrated Solution for Total Quality." *Intech*, ISA, Jan. 1990.

Shaw, W. T. *Computer Control of Batch Processes*. Cockeysville: EMC Controls, Inc., 1982.

Shewart, W. A. *Economic Control of Quality of Manufactured Product.* New York: Van Nostrand Company, Inc., 1931.

Shinskey, F. G. *Process-Control Systems*. New York: McGraw-Hill Book Company, 1979.

Southard, R. K. "Local Area Networks - An Overview, Part I" *Manufacturing Systems*, Carol Stream: Hitchcock Publishing Company, Jun. 1990.

Technical Publishing Company "Special Edition: Manufacturing Automation Protocol." *Control Engineering*, Barrington: Technical Publishing Company, Oct. 1985.

Trevathan, V. L. (Ed.) *A Guide to the Automation Body of Knowledge*. Research Triangle Park: ISA, 2006.

Walsh, S. "Perspectives on the Age of the Smart Machine." *Managing Automation,* Thomas Publishing, Dec. 1989.

Williams, T. J. *A Reference Model for Computer Integrated Manufacturing (CIM): A Description from the Viewpoint of Industrial Automation*. Research Triangle Park: ISA, 1989.

Zuboff, S. *In the Age of the Smart Machine*. New York: Basic Books, Inc., 1988.

Index

Wiley The manufacturer's authorized representative according to the EU
General Product Safety Regulation is Wiley-VCH GmbH, Boschstr. 12,
69469 Weinheim, Germany, e-mail: Product_Safety@wiley.com.

Printed and bound by CPI Group (UK) Ltd, Croydon, CR0 4YY

08/05/2026

02105976-0003